CAMBRIDGE TEXTS IN THE
HISTORY OF PHILOSOPHY

———

RENÉ DESCARTES
The World

T0296446

CAMBRIDGE TEXTS IN THE
HISTORY OF PHILOSOPHY

Series editors
KARL AMERIKS
Professor of Philosophy at the University of Notre Dame
DESMOND M. CLARKE
Professor of Philosophy at University College Cork

The main objective of Cambridge Texts in the History of Philosophy is to expand the range, variety and quality of texts in the history of philosophy which are available in English. The series includes texts by familiar names (such as Descartes and Kant) and also by less well-known authors. Wherever possible, texts are published in complete and unabridged form, and translations are specially commissioned for the series. Each volume contains a critical introduction together with a guide to further reading and any necessary glossaries and textual apparatus. The volumes are designed for student use at undergraduate and post-graduate level and will be of interest, not only to students of philosophy, but also to a wider audience of readers in the history of science, the history of theology and the history of ideas.

For a list of titles published in the series, please see end of book.

RENÉ DESCARTES

The World

and Other Writings

TRANSLATED AND EDITED BY

STEPHEN GAUKROGER

University of Sydney

CAMBRIDGE
UNIVERSITY PRESS

PUBLISHED BY THE PRESS SYNDICATE OF THE UNIVERSITY OF CAMBRIDGE
The Pitt Building, Trumpington Street, Cambridge CB2 1RP, United Kingdom

CAMBRIDGE UNIVERSITY PRESS
The Edinburgh Building, Cambridge CB2 2RU, United Kingdom
40 West 20th Street, New York, NY 10011–4211, USA
10 Stamford Road, Oakleigh, Melbourne 3166, Australia

© Cambridge University Press 1998

First published 1998

Typeset in Ehrhardt 11/13 [CP]

A catalogue record for this book is available from the British Library

Library of Congress cataloguing in publication data
Descartes, René, 1596–1650.
[Selections. English. 1998]
The world and other writings / René Descartes; translated and
edited by Stephen Gaukroger.
p. cm. – (Cambridge texts in the history of philosophy)
Includes bibliographical references and index.
ISBN 0 521 63158 0 (hardback). – ISBN 0 521 63646 9 (pbk.)
1. Optics – Early works to 1800. 1. Gaukroger, Stephen.
II. Title. III. Series.
QC353.D48213
500 – dc21 98–14404 CIP

ISBN 0 521 63158 0 hardback
ISBN 0 521 63646 9 paperback

Transferred to digital printing 2004

Contents

Acknowledgements

In translating and annotating *The World*, I have found a number of works of great value. In October 1996 a new edition of the texts with detailed annotations by Annie Bitbol-Hespériès and Jean-Pierre Verdet was published as *René Descartes: Le Monde, L'Homme* (Paris). Although this appeared after I had drafted the notes to this edition, I still found it of great use, as I did the annotations by Michael Mahoney to his translation of the *Treatise on Light* which appeared as *René Descartes: Le Monde, ou Traité de la Lumière* (New York, 1979). Above all, I am indebted to the copious and invaluable annotations by Thomas Hall in his translation of the *Treatise on Man*, which appeared as *René Descartes: Treatise of Man* (Cambridge, Mass., 1972). The other major work included in full here, the *Description of the Human Body*, has, so far as I know, not appeared in English in its entirety before.

I am indebted to Peter Anstey, Helen Irving, and John Sutton for their comments on parts of the translations, and to Desmond Clarke for his comments on the final draft.

Introduction

The origins of *The World*

The *Treatise on Light* and the *Treatise on Man* – which I shall refer to under the collective title *The World* – together constitute the most ambitious systematic project that Descartes ever undertook. Neither appeared in his lifetime. The first was published posthumously as *Le Monde* in 1664, the second two years earlier as *Renatus Descartes de Homine*. Both are parts of what is ostensibly a single work, and form the backbone of a single treatise. The text went through a number of redraftings, not just with respect to the detail of the arguments but also with respect to what should be included in the treatise, and the project included not only the *Treatise on Light* and the *Treatise on Man*, but also the material on the formation of colours in the *Meteors* and the material on geometrical optics in the *Dioptrics*, both subsequently published in 1637 along with the *Discourse* and the *Geometry*. I have included this material as appendices to the text of the *Treatise on Light*. There are also indications that Descartes had originally intended incorporating other material, including some work on music, for example, although this is not extant and may never have been developed in a systematic way.

The core doctrine at stake in *The World* is that of mechanism – above anything else, the doctrine that matter is completely inert – and Descartes' aim is to provide a mechanistic cosmology, resting on the basis of quantitative 'laws of nature', and a mechanistic physiology. Among the more fundamental things that he sets out to establish, four are of special significance and novelty. The first is that the stability of planetary orbits and the orbits of their moons can be accounted for on a mechanist basis if we envisage the planets being carried in a sea of fluid matter which takes the form of a vortex. The second is that the propagation of light from the Sun can be explained in terms of the centrifugal effects of its axial rotation. The third is that all vital functions can be accounted for mechanistically. And the fourth is that perceptual cognition can be accounted for, at least to a very large extent, in terms of a mechanistic psycho-physiology.

Descartes began *The World* in October 1629, and abandoned it, on hearing of the condemnation of Galileo,[1] at the end of 1633. From the account given in Part v of the *Discourse on Method*, it seems that the original project was designed to cover three topics: inanimate nature, animals and especially the human body, and the 'rational soul'. The descriptions of the first and second parts correspond closely to what we have in the *Treatise on Light* and the *Treatise on Man* respectively, but the third part of the project is not extant, and although in Part v of the *Discourse*, in describing *Treatise on Light*, Descartes says that after completing the *Treatise on Man* he 'described the rational soul', it may never have even been drafted at this time.

Descartes had earlier devoted some attention to at least some aspects of the three areas that he intended to pursue in *The World*, but with nothing like the breadth of vision.[2] Isaac Beeckman[3] had introduced Descartes to a micro-mechanical form of corpuscularian natural philosophy in late 1618/early 1619, and Descartes' early exercises in hydrostatics took the form of an attempt to explain macroscopic phenomena in micro-corpuscularian terms. This early work in statics provided him with his notions of 'action' and 'tendency to motion' (bodies hardly ever move in Descartes' mechanics: rather, they exhibit 'tendencies to motion', something encompassed by his general term 'action'), as well as providing him with a model of physical explanation in which one seeks to understand how physical processes are modified as one moves from one system of constraints to another, as opposed to the far more common seventeenth-century model of corpuscles moving in a void. Indeed, his early concern with explaining the behaviour of bodies in fluids is carried over in a

---

[1] Galileo Galilei (1564-1642) was the leading proponent of the heliocentric theory in the first part of the seventeenth century, his *Dialogue Concerning Two Chief World Systems* (1632) offering an elaborate defence of heliocentrism. It was quickly condemned by the Roman Inquisition, copies withdrawn, and Galileo, after recanting, was put under house arrest.

[2] For the details of Descartes' early intellectual development in the period before he began work on *Le Monde*, see chs. 1–6 of Stephen Gaukroger, *Descartes: An Intellectual Biography* (Oxford, 1995).

[3] Isaac Beeckman (1588–1637) was a Dutch engineer, physician, and educational administrator who pioneered the development of a purely mechanistic physical theory.

striking way into the *Treatise on Light*, where the motion of the planets results in large part from the motion of the fluid in which they are embedded. But Descartes' concern with physical questions at this time oscillated between extremely specific exercises, such as the explanation of free fall and Stevin's hydrostatic paradox in 1618–19, and very programmatic statements on universal method in the early 1620s. Although he had discovered the sine law of refraction some time in the mid- to late 1620s, and had investigated the physical basis for the law, before *The World* he seems to have had very little success in finding the right level at which to formulate a natural philosophy which had both real empirical content and offered a genuinely broad conceptual understanding of natural processes.

The subject of the *Treatise on Man* is animal physiology. Except for the question of perceptual cognition, there is no record of any general interest in anatomy and physiology before his move to the Netherlands at the end of 1628. Descartes had pursued perceptual cognition in an ingenious way in Rules 12–14 of the *Rules for the Direction of the Native Intelligence*, composed some time between 1626 and 1628. In these Rules he restricted his attention to the psycho-physiology of perceptual cognition, investigating the way in which sensed objects are represented by means of line lengths in the imagination. By this time Descartes had a firm understanding of geometrical optics, and a good basic understanding of faculty psychology, and his aim was to construe perceptual cognition along largely mechanist lines in terms that made no reference to the traditional 'vegetative' and 'sensitive' souls. By the time of his move to the Netherlands, however, we begin to find a more systematic interest in anatomy and physiology. He tells Mersenne[4] in a letter of 18 December 1629 (AT i. 102) that he has taken up the study of anatomy,[5] and during his first winter in Amsterdam he would visit the butcher daily to watch the slaughtering of cattle, and would take parts he intended to dissect back to his lodgings.[6] He seems to have kept up an interest in these topics throughout the period of composition of the first part of *The World*, and he continued work on the *Treatise on Man*, possibly making revisions to the manuscript, into the mid-1640s.

As regards the third part of the project, on the 'rational soul', although

---

[4] Marin Mersenne (1588–1648) was a pioneer in mechanist natural philosophy, musical theory and acoustics. His extensive correspondence forms the best guide to scientific thought in the years 1629-48. Descartes was one of his principal correspondents.

[5] Cf. Descartes to Mersenne, [20 February 1639], Adam and Tannery edition (abbreviated to AT hereafter) ii. 525.

[6] Descartes to Mersenne, [13 November 1639]; AT ii. 621.

we have nothing that Descartes may have written on this question from this time, we do know that, amongst other things, he was working on a treatise on metaphysics in 1629–30. He mentions that he had begun work on 'a little treatise' in a letter to Gibieuf[7] in July 1629, and a later letter to Mersenne indicates that this was a treatise on metaphysics in which he 'set out to prove the existence of God and of our souls when they are separate from the body, from which their immortality follows'.[8] These were the two traditional questions that Parisian philosophers of the 1620s had been concerned with, and the projected third part of *The World* may well have drawn on material in the abandoned treatise on metaphysics, although it should be noted that when Descartes later summarises his mechanistic physiology it will be in the *Passions of the Soul*, where it acts as a prelude to an account of the passions, rather than to a metaphysical discussion of the nature of mind.

As far as the provenance of *The World* is concerned, it had modest beginnings. In a letter of October 1629 Descartes wrote to Mersenne seeking fuller information on a particularly striking appearance of 'false' or 'multiple suns' – parhelia – observed by the astronomer Christoph Scheiner[9] at Frascati, just outside Rome, on 20 March. Descartes had become quite excited about the question, and, realising that the phenomenon bore a striking similarity to rainbows, dropped other projects, including the treatise on metaphysics.[10] He tells Mersenne that he has been working on meteorological questions generally, and that his interest has outgrown a concern merely to explain parhelia. He has resolved 'to write a small treatise on [meteorology] which will contain the explanation of the colours of the rainbow, which has given me more trouble than all the rest and, in general, all sublunary phenomena'. But this will be no ordinary treatise, 'for I have decided to exhibit it publicly as a sample of my Philosophy, and to hide behind the canvas to listen to what people will say about it' (AT i. 23). The topic is one of the best he could choose for this purpose, he tells Mersenne, and he promises to send him the manuscript for publication when it is complete, as he would prefer that it be published in Paris.

[7] Guillaume Gibieuf (c.1591–1650) was a French theologian who wrote on a number of metaphysical issues including freedom of the will.
[8] Descartes to Mersenne, 25 November 1630; AT i. 182.
[9] Christoph Scheiner (1575–1650) was a Jesuit mathematician and astronomer.
[10] In October Descartes tells Mersenne that he has had to interrupt what he has been working on, which is almost certainly a reference to the treatise on metaphysics. See Descartes to Mersenne, 8 October 1629; AT i. 23.

By November, the project has grown even further, and he writes to Mersenne on 13 November that:

> I should tell you that it will be more than a year before it is ready. For since I wrote to you a month ago, I have done nothing at all except sketch its argument, and instead of explaining a single phenomenon, I have decided to explain all natural phenomena, that is, the whole of physics. And the plan gives me more satisfaction than anything previously, for I think I have found a way of presenting my thoughts so that they satisfy everyone, and others will not be able to deny them. (AT i. 70)

The move from parhelia, first of all to meteorological phenomena, then to the whole of the physical world, is a huge one and it had taken shape in Descartes' mind over a period of no more than four months, between August and November 1629.

In Part 5 of the *Discourse*, Descartes sets out the details of the treatise he was working on in the period from mid-1629 to 1633. He writes:

> I tried to explain the principles in a Treatise which certain considerations prevented me from publishing, and I know of no better way of making them known than to set out here briefly what it contained. I had as my aim to include in it everything that I thought I knew before I wrote it about the nature of material things. But just as painters, not being able to represent all the different sides of a body equally well on a flat canvas, choose one of the main ones and set it facing the light, and shade the others so as to make them stand out only when viewed from the perspective of the chosen side; so too, fearing that I could not put everything I had in mind in my discourse, I undertook to expound fully only what I knew about light. Then, as the opportunity arose, I added something about the Sun and the fixed stars, because almost all of it comes from them; the heavens, because they transmit it; the planets, comets, and the earth, because they reflect light; and especially bodies on the earth, because they are coloured, or transparent, or luminous; and finally about man, because he observes these bodies. (AT vi. 41–2)

From 1629 to 1630, the problem that Descartes faced was that of building up his general knowledge of physics and related areas, sorting out what he should and should not concentrate upon, and finding a guiding thread by which to organise the argument of his treatise. The first and second problems took up a great deal of his time. In his letter to Mersenne of

15 April 1630 he complains that his 'work is going very slowly, because I take much more pleasure in acquiring knowledge than in putting into writing the little that I know' and that he is 'now studying chemistry and anatomy simultaneously'. But the third problem evidently gave him no less trouble. Later in the same letter he tells Mersenne that 'all these problems in physics that I told you I have taken on are all so interlinked and depend so much on one another that it is not possible for me to give a solution to one of them without giving a solution to all, and I cannot do that more quickly or more succinctly than in the treatise I am writing'. And in a letter to Mersenne of 23 December of that year he tells him that he has 'countless different things to consider all at once' and that he is trying to find some 'basis on which to give a true account without doing violence to anyone's imagination or shocking received opinion' (AT i. 194).

As for the order of composition, we know that Descartes worked on the *Treatise on Light* between 1630 and 1632. There is some evidence that it was proceeding in fits and starts in 1630, and that work on it was taking its toll on Descartes. In a letter of 15 April, for example, he asks Mersenne not to confirm to anyone that he is writing his treatise on physics but rather to give them the impression that he is not, for 'I swear that if I had not already told people that I planned to do so, with the result that they would say that I had not been able to carry out my plan, I would never undertake the task.' On the assumption that he wrote the chapters in the order in which they appear in the extant draft, he was up to chapter 5 by the end of February 1630 and had completed the material for chapters 6 to 8 in the first three months of 1632, the remaining chapters being drafted between then and late 1632. From late 1632 he concentrated on the *Treatise on Man*, and he had already done a considerable amount of work in physiology by this stage. At the end of 1632 he tells Mersenne that he will

> speak more about man than I had intended to before, because I shall try to explain all of his principal functions. I have already written about those that pertain to life, such as the digestion of food, the beating of the pulse, the distribution of nutrients etc., and the five senses. Now I am dissecting the heads of different animals in order to explain what imagination, memory etc., consist of. I have seen the book *De motu cordis* [of Harvey[11]] of which you spoke

---

[11] William Harvey (1578–1657) was an English physician and physiologist who discovered the circulation of the blood. His account of circulation, and the pumping action of the heart, was set out in *De motu cordis* (1628).

to me earlier, and find I differ only a little from his view, which I came across only after I had finished writing about this matter. (AT i. 263)

From this time until mid-1633 he appears to have devoted himself to physiology.

The *Treatise on Light*

The first five chapters of the *Treatise on Light* form a kind of introduction, suggesting that matter and motion are sufficient to explain natural phenomena, and proceeding to set out the theory that the material world consists exclusively of matter (in particular, does not have any empty regions), and that this matter can be considered as comprising three sizes of corpuscle. The defence of mechanism offered starts off in the first three chapters as a very general and intuitive one, appealing to common-sense examples, while the remaining chapters of the first part shift to a more contentious version of mechanism, as Descartes moves from a considera- tion of the nature of liquidity and hardness to a micro-corpuscular theory of elements and a rejection of an inter-corpuscular void. Chapters 6 to 14 then use this micro-corpuscular theory of matter, combined with a number of laws describing the motion of the corpuscles, to set out a mech- anistic cosmology which includes both a celestial physics and an account of the nature and properties of light. The text ends abruptly with an un- finished chapter 15.[12]

In the first chapter Descartes shows that our perceptual images need not resemble what they represent. What he is attacking here is the prevalent Aristotelian view that the veridicality of our perceptual images of the world lies in their ability to resemble the objects perceived. Descartes provides an example to show that a sensation need not resemble the cause of that sensation. But his account raises two deeper issues. First, he introduces a positive account of visual cognition according to which the way in which perceptual images represent the object perceived is modelled not on pictorial representation (as it had been in the *Rules*, for example, where the perceptual image takes the form of lines 'etched' on a two-dimensional surface), but on verbal understanding, so that our attention is drawn to the phenomenon of what might be called visual understanding, something that

---

[12] The original, presumably complete, manuscript seems to have consisted in 18 chapters: see AT xi. iii–iv. We have no indication what the content of the missing chapters was.

involves an irreducible element of interpretation on our part. Secondly, the thrust of his negative argument against the resemblance theory goes beyond the claim that the world may be different from our perceptual image of it, and what Descartes is really trying to steer us towards is the idea that our perceptual image may not even be a guide to how the world is. In particular, he suggests that light may be 'different in objects from what it is in our eyes'.

In the second chapter, he starts out on the task of establishing this. Turning directly to the nature of light, he points out that there are only two sorts of bodies in which light is found, namely the stars, and flame or fire, the latter being the more familiar and hence the best starting-point. The aim is to show how a macroscopic phenomenon can be accounted for plausibly in micro-corpuscularian terms, and fire is a good example for Descartes' purposes. We can see that the fire moves the subtler parts of the wood and separates them from one another, transforming them into fire, air, and smoke, and leaving the grosser pieces as ashes. All we need to postulate in order to account for the burning process is the motion of parts of the wood resulting in the separation of the subtle parts from the gross parts.

In the course of this discussion, Descartes tells us that he is not concerned with the direction of motion, making a sharp distinction between speed and direction. Now since, when something moves, it always moves in a direction, motion would appear to have both speed and direction, these being two inseparable components of the same thing. But Descartes sees matters differently. For him, the power by which something moves and the power which determines its motion as being in one direction rather than another are different powers. In his *Dioptrics*, to which he refers us here, he gives the example of a tennis ball being reflected off a hard surface. The thrust of the argument is that, because the tennis ball and the surface are inelastic, if force and direction of motion were the same thing then the ball would first have to stop before it changed direction, and if it stopped a new cause would be needed for it to move again. But there is no such new cause available: therefore, its force is not affected in the impact, only the direction of its motion, which is changed. The basic distinction that he wants to draw is between the power by which something moves and its 'determination', the latter being something that depends on the force or speed of the body, and which directs that speed or force. The geometrical configuration of other bodies can alter this determination and Descartes goes on to tell us that the actual path of a moving body is determined by each part moving 'in the manner made least difficult for it' by surrounding bodies.

At the beginning of chapter 3, Descartes draws attention to the prevalence of change in nature, but he argues that the total amount of motion in the universe is conserved, although this motion may be redistributed among bodies. His account of the difference between hard and fluid bodies in chapter 3 forms a bridge between a very general statement of the mechanist position, most of which would have been common ground to mechanists, and a specific version of micro-corpuscularianism which was both more distinctive and more contentious. The general principle from which Descartes works is that, given that all bodies can be divided into very small parts, a force is required to separate these parts if they are stationary with respect to one another, for they will not move apart of their own accord. If the very small parts of which the body is constituted are all at rest with respect to one another then it will require significant force to separate them, but if they are moving with respect to one another then they will separate from one another at a rate which may even be greater than that which one could achieve by applying a force oneself. The former bodies are what we call solids, the latter what we call fluids, and in the extreme cases they form the ends of a spectrum on which all bodies can be ranked, with rigid solids at one terminus and extremely fluid bodies at the other. This ranking on a spectrum of fluidity provides the basis for Descartes' theory of matter, for it enables him to reduce the properties of matter to the rate at which its parts move with respect to one another. At the extreme fluid end of the spectrum comes, not air as one might expect, but fire, whose parts are the most obviously agitated, and whose degree of corpuscular agitation is such that it renders other bodies fluid.

The discussion of the nature of air in chapter 4 opens with the question of the existence of imperceptible bodies. Descartes tells us he is clearing away a prejudice which we have from childhood, that the only bodies that exist are those that can be sensed, and that air is so faintly sensible that it cannot be as material or solid as those we perceive more clearly. All bodies, whether fluid or solid, are made from the one kind of matter. Descartes argues that the degree of fluidity of a body cannot be proportional to the amount of vacuum that exists between its constituent parts, trying to establish that there must be more space between the parts of a solid than between those of a liquid, because the moving parts of a liquid 'can much more easily press and arrange themselves against one another' than can the parts of a solid. His main conclusion is that if there is a vacuum anywhere it cannot be in fluids but must be in solid bodies, and he is more concerned to make sure that we accept that there are no interstitial vacua in fluids than to show the absence of such vacua in solids. This is because his account of the basic structure of the universe effectively subsumes it under fluid mechanics, and hence his interest is really in fluids. This begins to become evident in the subsequent discussion, where the question of the non-existence of a void is discussed in terms of the motion of fluids, and it becomes part of a question in fluid mechanics. In particular, the question arises of how bodies can move at all if there is no empty space for them to move into, and the answer Descartes gives is that 'all the motions that occur in the world are in some way circular'. With circular motion, matter could move in a plenum by means of a large-scale displacement: a region of matter will then be able to move when contiguous matter in the direction of its motion, and contiguous matter in the opposite direction, also move in the direction of its motion, and when the same conditions hold for these contiguous pieces of matter, so that in the end a continuous loop or ring of matter is displaced.

The doctrine of elements immediately follows the account of circular translation. He invokes only three elements – fire, air, and earth. This is to be explained by the fact that Descartes is writing a treatise on *light*. At an intuitive level, three kinds of process are involved, namely, the production of light, the transmission of light, and its reflection and refraction. Descartes' model of light is one drawn from fluid mechanics: it is something that acts by means of mechanical pressure, and what needs to be explained is how this mechanical pressure is generated in the first place, how it is propagated, and why light so construed behaves in particular geometrically defined ways when it encounters opaque and transparent bodies. Light is generated by fiery bodies, transmitted through the air, and is refracted and reflected by terrestrious bodies. The traditional elements of fire, air, and earth have, then, a cosmological analogue. These three elements are for Descartes simply three different sizes of corpuscle: very fine, fine, and gross respectively. They are the kinds of matter Descartes believes one needs for a physical theory of light, and become unashamedly hypothetical by the end of the chapter, where Descartes tells us that he is going to 'wrap his discourse up in the cloak of a fable'.

Chapter 6 begins with Descartes' construction of a hypothetical world on the basis of the theory of matter set out in the first five chapters. The ultimate aim is to show that a world constructed in this manner, one without forms or qualities, is indistinguishable from the actual world. The traditional forms and qualities are excluded because they could not form part of a properly mechanist explanation. The task of the first five chapters has been to set out the kind of entities and properties that he wants to invoke in his account, and he has prepared the ground by trying to show that they have the requisite qualities of clarity and evidence. If we strip the world of the traditional forms and qualities, what we would be left with would, in Descartes' view, be its genuine properties. His new world is to be conceived as 'a real, perfectly solid body which uniformly fills the entire length, breadth, and depth of the great space at the centre of which we have halted our thought'. This perfectly solid body is 'solid' in the sense of being full and voidless, and it is divided into parts distinguished simply by their different motions. At the first instant of creation, God provides the parts with different motions, and after that He does not intervene supernaturally to regulate their motions. Rather, these motions are regulated by laws of nature which Descartes now sets out.

These laws of nature are designed to describe the collisions of corpuscles. In imagining such collisions, it is tempting to picture them in terms of atoms colliding in a void, but we must exercise care in allowing ourselves to think in these simplified terms, for we naturally think of atoms moving in a void as continuing for long stretches without collision, whereas for Descartes there is constant collision. This is important because the counterfactual situation in which a body moves in the absence of external constraints is not so immediately relevant to Descartes' analysis as it would be to a straightforwardly atomistic account, where the obvious way to proceed would be from the simple case of unconstrained motion to how the motion is changed by various constraints. This is the essence of the kinematic approach,[13] but it is far from clear that Descartes' approach is kinematic. His model seems rather to be taken from hydrostatics, and the point seems to be not so much to analyse the behaviour of a body under various kinds of constraint in terms of how it behaves when not under constraint, but rather to account for what happens when a body moves from one system of constraints to another, where the constraints that Descartes is interested in are collisions.

The three laws of nature that Descartes provides are designed to

---

[13] See, for example, Galileo's brilliant kinematic analysis of falling bodies in terms of motion in a void, in the second half of the 'First Day' of his *Two New Sciences*, published in 1638.

describe the behaviour of bodies in collision. They deal quite separately with the power of moving and the determination of a body. The first law tells us that a body conserves its motion except in collision, when, the second law tells us, the total motion of the colliding bodies is conserved but may be redistributed amongst them. It is left to the third law to tell us about direction, and according to this law, because a body's tendency to move is instantaneous, this tendency to move can only be rectilinear, because only rectilinear motion can be determined in an instant: 'only motion in a straight line is entirely simple and has a nature which may be grasped wholly in an instant'. Motion in a circle or some other path would require us to consider 'at least two of its instants, or rather two of its parts, and the relation between them'. What path the body will actually take, however, will be a function of the collisions to which it is subject.

The first law states that certain states of bodies are conserved: they will remain unchanged unless something acts to change them. Among these are a body's size, shape, its position if it is at rest, and also its motion, for once a body has begun to move, 'it will always continue in its motion with an equal force until others stop or retard it'. This rule of conservation of state has always been considered to hold for the first three items, and many others, Descartes tells us, but not for the last, 'which is, however, the thing I most expressly wish to include in it'. In defence of the first law, Descartes spells out the conception of motion that it employs and contrasts this with the Aristotelian conception. His suggestion is that motion is simply to be equated with change of place or translation.

The second law of motion is a law of the conservation of motion (or perhaps a law of conservation of the total 'force of motion') in collisions. In its defence, Descartes points to its advantages over the traditional accounts of continued projectile motion. Aristotelians were in disagreement amongst themselves about how to account for the continued motion of projectiles, and their accounts were premised upon a distinction between terrestrial and celestial motions. Descartes changes the question, so that it now becomes that of explaining why the motion of the projectile decays rather than why it continues to move, and the answer he provides is the air's 'resistance'.

When he sets out the second law he talks about motion being conserved, but in subsequent elaboration he reformulates it in terms of conservation of 'force of motion'. Because of the problems in separating out what exactly is physical and what is divine in Descartes' account of causation and force, it is difficult to say whether causation is something physical, or whether it has both a divine manifestation and a physical manifestation in the form of force of motion, or whether force of motion is a physical expression of something that is non-physical.[14] But whichever of these we opt for, motion is conserved because force of motion is conserved, and force of motion in some way expresses or manifests God's causal activity. It is ultimately because causation is conserved – a conservation that Descartes puts in terms of God's immutability – that motion is conserved. `

Whereas the first two laws deal with the power of motion, the third deals with what Descartes regards as a separate issue: the direction of motion. It asserts that, whatever the path of a moving body, its tendency to motion, or *action*, is always rectilinear. The evidence presented for this is (i) that a stone released from a sling will not continue to move in a circle but will fly off along the tangent to the circle, and (ii) while in the sling the stone will exert a force away from the centre causing the string to stretch, showing 'that it goes around only under constraint'. But there is a notorious discrepancy in Descartes' account here. The trouble is that while the third law as stated in chapter 7 would seem to establish the uniqueness of recti-linear motion as an inertial motion, when he elaborates further on the law in chapter 13 he apparently counts a circular component in the motion of the stone as inertial as well. Why, after giving a clear statement of rectilinear inertia and providing an explanation of why rectilinear motion is the only inertial motion in terms of its 'simplicity', does he appear to blatantly contradict this? There are two complementary answers to this question, I believe, and both derive from Descartes' attempt to use the hydrostatic model in his physical theory. The first is that a statement of a principle of inertia does not seem to be the main point of the exercise. He does not seem particularly concerned to specify how a body behaves in the absence of forces, for example, because the bodies he deals with always move within a system of constraints, just as in statics: the aim is to understand the instantaneous collisions of non-elastic bodies. One does not ask what would happen if the forces were removed, because the understanding of the action of these forces is the point of the exercise. The second is that what Descartes is concerned with in chapter 13 is not so much circular inertia as circular equilibrium, namely, the idea that a body moves in a continuous

[14] See M. Gueroult, 'The metaphysics and physics of force in Descartes', in S. Gaukroger (ed.), *Descartes: Philosophy, Mathematics and Physics* (Brighton, 1980), 196–229, *passim*, and Gabbey, ibid., 234–9.

circular orbit because the forces acting upon it are exactly balanced, so that the net force is zero. The confusion arises because Descartes slides between this static notion of equilibrium (which involves the extremely problematic assumption that some motions are dynamically unbalanced) and the dynamic notion of inertia.

Chapters 8 to 12, using the theory of matter and laws of nature which have now been elaborated, set out the details of a heliocentric cosmology in the form of an account of a hypothetical 'new world', from the formation of the Sun and the stars (ch. 8), the planets and comets (ch. 9), the Earth and the Moon (ch. 10), and finally weight or gravity (ch. 11) and the tides (ch. 12). The key to this whole cosmology is Descartes' account of vortices. Because the universe is a plenum, for any part of it to move it is necessary that other parts of it move, and the simplest form of motion which takes the form of displacement is going to be a closed curve, although we have no reason to think that the universe turns around a single centre: rather, we may imagine different centres of motion. The matter revolving furthest away will be the largest or most agitated because it will describe the greatest circles, owing to its greater capacity to realise its inclination to continue motion in a straight line. Whatever differences in size and agitation we may imagine there to have been in the early stages of the universe, however, except for the large clumps of third element we can imagine that the constant motion and collision caused the difference in sizes of matter to be reduced as 'the larger pieces had to break and divide in order to pass through the same places as those that preceded them'. Similarly, differences in shape gradually disappear as repeated collisions smooth off the edges and all matter (of the second element) becomes rounded. Some pieces of matter are sufficiently large to avoid being broken down and rounded off in this way: these are what Descartes refers to as the third element, and such pieces of matter form the planets and the comets. Finally, the collisions yield very small parts of matter, which accommodate themselves to the space available so that a void is not formed but this first element is formed in a greater quantity than is needed simply to fill in the spaces between pieces of second and third element, and the excess naturally moves towards the centre because the second element has a greater centrifugal tendency to move to the periphery, leaving the centre the only place for the first element to settle. There it forms perfectly fluid bodies which rotate at a greater rate than surrounding bodies and exclude fine matter from their surfaces. These concentrations of first element in the form of fluid, round bodies at the centre of each system are suns, and the pushing action at their surfaces is 'what we shall take to be light'.

The universe, as Descartes represents it, consists then of an indefinite number of contiguous vortices, each with a sun or star at the centre, and planets revolving around this centre carried along by the second element. Occasionally, however, planets may be moving so quickly as to be carried outside the solar system altogether: then they become comets. Descartes describes the difference between the paths of planets and comets in terms of an engaging analogy with bodies being carried along by rivers, the latter being like bodies that will have enough mass and speed to be carried from one river to another when rivers meet, whereas the former will just be carried along by the flow of their own river. Planets eventually enter into stable orbits, the less massive they are the closer to the centre, and once in this orbit they are simply carried along by the celestial fluid in which they are embedded. The stability of their orbits arises because, once a planet has attained a stable orbit, if it were to move inward it would immediately meet smaller and faster corpuscles of second element which would push it outward, and if it were to move outward, it would immediately meet larger corpuscles which would slow it down and make it move inward again.

This accounts for the motions of comets and the motion of planets proper around the Sun, and Descartes now moves on to explain the motions of planetary satellites and the diurnal rotation of a planet like the Earth. The celestial matter in which the Earth is embedded moves faster at one side of the planet than at the other, and this gives the Earth a 'spin' or rotation, which in turn sets up a centrifugal effect, creating a small vortex around itself, in which the Moon is carried. Turning next to consider what the weight (*pesanteur*) of the Earth consists in, Descartes rejects the idea of weight as an intrinsic property. In earlier writings he had defined weight in functional terms as 'the force of motion by which a body is impelled in the first instant of its motion' (AT x. 68). It is not surprising, therefore, that he has no hesitation in offering a similar account here.

Finally, the phenomenon of the tides is explained using the same materials. Direct evidence for the orbital and rotational motion of the Earth was not available in the seventeenth century, but the tides, which are difficult to explain on the assumption of a non-rotating Earth, do offer indirect evidence. Tides are a very complicated phenomenon, however, involving daily, half-monthly, monthly, and half-yearly cycles. Descartes was especially pleased with his account and wrote to Mersenne at the time that accounting for the tides had given him a great deal of trouble, and that while he was not happy with all the details, he did not doubt the success of his account (AT i. 261). And although he will revise it over the next ten years, he will not alter its fundamentals. Indeed, the theory of the tides is really the first genuinely quantitative ingredient in the *Treatise on Light*, but the fact that the earlier material is not quantitative should not blind us to the significance of Descartes' success in presenting a thoroughly mechanist cosmology which takes as its foundations a strictly mechanist conception of matter and the three laws of motion. The *Treatise on Light* presents a fully mechanist alternative to Aristotelian systems, one which effectively derives heliocentrism from first principles, which offers a novel and apparently viable conception of matter, and which formulates fundamental laws of motion – laws that are clearly open to quantitative elaboration. But the jewel in the crown of *Treatise on Light* is the theory of light set out in the last three chapters, for, especially if we read these together with Descartes' general work in optics at this time, we have an empirical, quantitative account of a physical question whose explanation derives directly from his mechanist cosmology.

Descartes' purpose in the last three chapters is to show how the behaviour of light rays can ultimately be explained in terms of his theory of the nature of matter and the three laws of motion. Indeed, the theory of matter turns out to be motivated directly by the requirements of Descartes' physical optics, for the first element makes up those bodies that produce light, namely suns and stars; the second element makes up the medium in which light is propagated, namely the celestial fluid; and those bodies that refract and reflect light, such as the planets, are made up from the third element. Moreover, it is the laws of motion that underpin and explain the laws of refraction and reflection of light, and the accounts of phenomena such as the rainbow and parhelia that are based on these.

The laws of motion show us that, given the rotation of the Sun and the matter around it, there is a radial pressure which spreads outwards from the Sun along straight lines from its centre. This pressure is manifested as 'a trembling movement', a property which is 'very suitable for light'. Indeed, the inhabitants of Descartes' proposed new world 'have a nature such that, when their eyes are pushed in this way, they will have a sensation which is just like the one we have of light'. The question that Descartes now poses is whether this model accounts for the known properties of light. Setting out twelve 'principal' properties of light which a theory of light must account for he proceeds to show that his account is not only compatible with all of these, but can actually explain them.

Descartes' achievement in the *Treatise on Light* is twofold. In the first place, his vortex theory explains the stability of planetary orbits in a way that presents an intuitively plausible picture of orbital motion which requires no mysterious forces acting at a distance: the rapid rotation of the Sun at the centre of our solar system, through its resultant centrifugal force, causes the 'pool' of second matter to swirl around it, holding planets in orbits as a whirlpool holds bodies in a circular motion around it. Moreover, it explains this motion in terms of fundamental quantifiable physical notions, namely centrifugal force and the rectilinear tendencies of moving matter. In other words, the heliocentric theory is derived from a very simple theory of matter, three laws of motion, and the notion of a centrifugal force. Secondly, this account also enables Descartes to account for all the known principal properties of light, thereby providing a physical basis for the geometrical optics that he had pursued so fruitfully in the 1620s.

The *Treatise on Man*

Just as the strategy behind the *Treatise on Light* was to construct an 'imaginary world' out of the basic materials supplied by mechanism, and then show that such a world is indistinguishable from the real one, so too the *Treatise on Man* begins with an imaginary mechanistic world, the aim being to show how a physiology can be constructed out of it which is indistinguishable from real animal physiology. The physiology he describes is not original, being derived from a number of sources including Hippocrates, Galen, Scholastic commentators on biology and medicine, especially the Coimbra commentators, and various sixteenth- and seventeenth-century writers on biological and medical topics.[15] The originality comes in the attempt to show how such a physiology can be modelled mechanistically. In particular, various functions had traditionally been ascribed to qualitatively different 'souls': digestion, movement of the

[15] For details see Annie Bitbol-Hespériès, *Le Principe de vie chez Descartes* (Paris, 1990). Hippocrates (c.460–c.377 BC) is widely regarded as the founder of medicine. Galen (c.130–c.210) is the most important author on anatomy and physiology before Versalius and Harvey respectively. The Coimbra commentators were a group of Jesuit philosophers based at Coimbra in Portugal who produced huge commentaries on Aristotle in the last decades of the sixteenth and early seventeenth centuries. Their works were standard in Jesuit colleges, and Descartes was taught from them.

blood, nutrition, growth, reproduction and respiration to the 'vegetative soul'; perception, appetites and animal motion to the 'sensitive soul'. Descartes sets out to show how we need postulate no souls at all for these organic processes, that all that is needed is the right kind of mechanical explanation.

In Part 1, for example, the digestion of food is described in a mixture of mechanical and chemical terms, and the cause of the circulation of the blood is put down to the production of heat in the heart, the thermogenetic processes causing pressure in the arteries. The blood carries animal spirits, which are separated out from the blood by a simple filtration process and enter the brain through the pineal gland. In Part 2, he sets out how the nervous system works by means of the animal spirits, which enter the nerves and change the shape of the muscles. It is worth remembering in this context that Descartes' mechanistic model is not that of a clock, but one of hydraulic systems, such as those that worked the fountains and moving statues in the gardens of Saint-Germain. Just as in the *Treatise on Light*, where bodies are carried along in fluids, so here the kind of image Descartes' model conveys is that of fluids being pushed through tubes, not wheels working cogs, and this has a much more intuitively 'organic' feel, something that Descartes' critics have often overlooked when assessing the general plausibility of his account.

A crucial discussion in Part 2 is that of the action of the pineal gland, which is also responsible for the discharge of the animal spirits to the muscles via the nerves. Take the case of an animal spotting a predator and escaping. Physiologically, what happens is that external stimuli – smells and visual stimulation – displace the peripheral ends of the nerve fibres in the nose and eyes, and structural isomorphs of the sense impressions are transmitted to the brain, unified into a single isomorph in the 'common sense' (which unifies isomorphs from the various senses into a unitary image), and form an 'idea' – a change in the pattern formed by the animal spirits on the surface of the pineal gland. Such a changed pattern results in changes in the outflow of spirits to the nerves. At the muscle, a small influx of spirit from the nerve causes the spirits already there to open a valve into its antagonist. Spirits then flow from the antagonist which causes it to relax, as well as causing the first muscle to contract. Escape from the predator is thereby facilitated. Note, incidentally (a fact overlooked by very many commentators), that this is *not* an account of reflex action, which is described at the end of Part 2, for reflex action is a more primitive operation which does not even involve the pineal gland since it does not require a representation of the stimulus but a direct response: the reflex arc (see fig. 38) passes through what Descartes refers to as a 'cavity' (labelled F), a term which he never uses for the pineal gland and which almost certainly refers to one of the cerebral ventricles.

In Part 3, which deals with sense perception, we are offered quite a sophisticated account of distance vision. Here Descartes deals with a particularly pressing problem for a mechanistic account of vision. The mechanist allows only contact actions, so the surface of the eye only has contact with the light corpuscles that strike it, but such corpuscles cannot carry information about their origins, for example, about the distance of their source. How, then, is distance perception possible? Descartes' ingenious solution is to suggest that our cognitive apparatuses (not our minds, for animals are capable of distance perception and they lack minds in the strict sense) are able to operate with an innate 'natural geometry' by which one can judge the distance of an object in virtue of knowing the distance between the eyes and the angle at which light corpuscles strike the eyes, this giving us the base angles and base length of a triangle, from which we can gauge the distance of the apex from the base by elementary trigonometry.

Part 4 is concerned with internal psycho-physiological operations, and here we are presented with a range of accounts of very different degrees of sophistication and plausibility. At one extreme is the account of personality traits such as generosity and liberality, which are put down to an abundance of animal spirits. At the other is an account of memory which mirrors the concerns of his account of perceptual cognition. In the latter case, Descartes is concerned to argue that perception does not involve resemblance, but it does involve representation. He now applies these considerations to memory, showing that memories need not resemble the event of which they are the memory: they need only encode the information in such a way that we can bring that event to mind in the absence of any external stimulus. Pineal patterns do not have to be kept in the same form between experiencing and remembering, and this has ramifications for the question of how memories are stored and retrieved, which steers Descartes' account, which is essentially a dispositional account, in a completely different direction from that of his predecessors and contemporaries, who were concerned with the question of the localisation of memory.[16]

[16] For details see John Sutton, *Philosophy and Memory Traces* (Cambridge, 1998).

The *Treatise on Man* reads like a complete treatise – the last sentences sum up the main thrust of the *Treatise* and have the air of a conclusion – but there are omissions, and we might have expected the argument that the mechanical devices constructed are indistinguishable in their operations from animal physiology to have been put, and a transition made to Part III, that is, to the case of human beings. The psycho-physiology (as just described) is regulated by a mind – most importantly the ability to make judgements and exercise free will – in the case of human beings, and this makes a crucial difference to the nature of their cognitive states, and it is a great pity that Descartes does not go on to spell out the nature of this difference. The situation is complicated by the fact that the *Treatise on Man* was worked on into the 1640s, however, for it may have become independent of the originally planned third part of the project.

The abandonment of *The World*

At the time that Descartes began working on *The World*, Galileo was putting the finishing touches to his *Dialogue Concerning the Two Chief World Systems*, to which he had devoted much of his time between 1624 and 1630. The work was withdrawn shortly after its publication in Florence in March 1632, however, and it was condemned by the Roman Inquisition on 23 July 1633.

The *Dialogue* provided physical evidence both for the Earth's diurnal rotation, in the tides, and for its annual orbital motion, in cyclical change in sunspot paths. It also provided a detailed and ingenious account of why our perceptual experience apparently does not accord with the Earth's motion, in the principle of the relativity of motion. Although Galileo was powerfully connected and was widely celebrated for his discovery of the moons of Jupiter in 1610, he had been warned of his responsibility to treat the motion of the Earth hypothetically by the Florentine Inquisition as early as 1616. This earlier condemnation, as well as that of 1633, focused on the question of the physical reality of the Copernican hypothesis. A core issue in dispute in both the 1616 and 1633 condemnations of Copernicanism was whether the heliocentric theory was 'a matter of faith and morals' which the second decree of the Council of Trent had given the Church the sole power to decide.[17] Galileo and his defenders denied that it was, maintaining that the motion of the Earth and the stability of the

---

[17] For details see Richard J. Blackwell, *Galileo, Bellarmine, and the Bible* (Notre Dame, 1991).

Sun were covered by the first criterion in Melchior Cano's handbook of post-Tridentine orthodoxy, *De locis theologicis*, namely that when the authority of the Church Fathers 'pertains to the faculties contained within the natural light of reason, it does not provide certain arguments but only arguments as strong as reason itself when in agreement with nature'. Opponents of Galileo argued that the case was covered by different criteria, such as the sixth, which states that the Church Fathers, if they agree on something, 'cannot err on dogmas of the faith'. In the 1633 condemnation, the latter interpretation was effectively established, and this meant that the physical motion of the Earth could not be established by natural-philosophical means. Thus the kind of arguments that Galileo had offered in the *Dialogue* not only had no standing in deciding the issue, but also the kind of arguments that Descartes had offered in the *Treatise on Light* had no such standing either.

At the end of November 1633, Descartes wrote to Mersenne:

> I had intended to send you *Le Monde* as a New Year gift . . . but in the meantime I tried to find out in Leiden and Amsterdam whether Galileo's *World System* was available, as I thought I had heard that it was published in Italy last year. I was told that it had indeed been published, but that all copies had been burned at Rome, and that Galileo had been convicted and fined. I was so surprised by this that I nearly decided to burn all my papers, or at least let no one see them. For I couldn't imagine that he – an Italian and, I believe, in favour with the Pope – could have been made a criminal, just because he tried, as he certainly did, to establish that the earth moves . . . I must admit that if this view is false, then so too are the entire foundations of my philosophy, for it can be demonstrated from them quite clearly. And it is such an integral part of my treatise that I couldn't remove it without making the whole work defective. But for all that, I wouldn't want to publish a discourse which had a single word that the Church disapproved of; so I prefer to suppress it rather than publish it in a mutilated form. (AT i. 270–1)

Descartes was clearly devastated by the condemnation, and he abandoned any attempt to publish *The World* as a result.[18] The outcome of this

---

[18] That it was indeed the condemnation of Galileo that prevented publication, and not his corpuscularianism or animal automatism, as have occasionally been suggested, is clear from the letter to Mersenne just quoted, and also from his request to Mersenne to tell Naudé that the only thing stopping him publishing his physics was the prohibition on advocating the physical reality of the Earth's motion. Descartes to Mersenne, December 1649; AT iii. 258.

crisis was a new direction in his work. Although he does not abandon interest in natural philosophy, and to the end of his life continues to think it has been his most important contribution,[19] this interest in it is now confined largely to polemics and systematisation, and above all to the legitimation of a mechanist natural philosophy by metaphysical and epistemological means,[20] a completely different enterprise from that pursued in *The World*.

The optical material in *The Treatise on Light* was to appear in the *Dioptrics* and *Meteors* (both 1637), shorn of all contentious cosmological material; the cosmological material, buttressed by metaphysical arguments and with a protective hypothesis which purported to show that all motion is relative (and so allowing one to hedge one's bets on the physical reality of a heliocentric system) appeared in the *Principles of Philosophy* (1644). As regards the physiology of the *Treatise on Man*, very basic outlines of the physiology were presented at the beginning of the *Passions of the Soul* (1649), but, other than that, nothing appeared in Descartes' lifetime. There are manuscript notes on anatomy, physiology, and embryology dating from the 1630s and early 1640s, and a much more significant piece called *Description of the Human Body* dating from the winter of 1647–8. The first three parts of this latter work update the *Treatise on Man*, and Descartes rejects Harvey's account of the pump action of the heart, preferring his own thermogenetic account as an explanation of the cause of circulation. In the fourth and fifth parts, new material on the development of the embryo is introduced. This is very important material, as the kinds of process involved in the formation of the foetus are far more intractable, from the point of view of a mechanist physiology, than basic adult physiology, and Descartes has to account for these processes purely in terms of his theory of matter. Indeed, the two crucial problem areas for a mechanist physiology are psycho-physiology and the formation of the foetus: both areas are constituted by what are apparently goal-directed activities – cognition in the one case, and growth from a relatively undifferentiated small mass of material to a complex organism in the other – and they offer an immense challenge to a mechanist account.

---

[19] See the passage in the conversation with Burman, given at AT v. 165, where Descartes warns against devoting too much time to metaphysical questions, especially to his *Meditationes*. These are just preparation for the main questions, which concern physical and observable things. Cf. Descartes to Elizabeth, 28 June 1643; AT iii. 695.

[20] This shift in direction of Descartes' thought is discussed in the detail in chs. 8 and 9 of Gaukroger, *Descartes: An Intellectual Biography*.

Descartes' treatment of the first is a triumph of sophistication and ingenuity; his treatment of the second is far more tentative, and much less successful, but he does make a sustained effort to come to terms with the problems.

The material on the soul in the projected third part of *The World*, which I have indicated was probably not even drafted as such at the time, but which would almost certainly have drawn on the lost 'Treatise on Metaphysics', took on a life of its own, so to speak, in the legitimatory epistemological–metaphysical project of the post-*Le Monde* period. One can properly raise the question whether this project yields its fruit in the radical dualism of the *Meditations* or in the rather more naturalistic account of mind in the *Passions of the Soul* (which begins with a summary of the *Treatise on Man*). The latter is suggested by structural reasons. The *Treatise on Light*, which deals with physics and cosmology, was followed by the *Treatise on Man*, which dealt with animal physiology, and was to be followed by a treatise on the soul. The four books of the *Principles of Philosophy* follow the structure of the *Treatise on Light*, albeit now the account is formulated in the context of a foundationalist metaphysics, in which dualism plays an integral role. The projected fifth book of the *Principles* was on 'living beings' and I think there can be no doubt that the *Description of the Human Body* is the draft material for that part. The *Passions of the Soul*, which offers a relatively naturalistic account of affective states,[21] employing Descartes' idea of a 'substantial union of mind and body' rather than the radical dualism of his foundational projects, seems to be the projected sixth part, on 'the soul'.

The World provides us with an alternative to the *Meditations* as an entry into Descartes' thought. Whereas in the *Meditations* we are led to natural philosophy through the sceptically driven epistemology on which Descartes grounds a metaphysically formulated natural philosophy, in *The World* we are offered a more direct access to the whole of natural philosophy, from cosmology to cardiology to the psycho-physiology of perception. If we take the latter route we are in a better position to assess the role played by, and any benefits to be derived from, the epistemological and metaphysical underpinning that Descartes provides for his natural philosophy in his later writings.

---

[21] The account is quite at odds with the radical dualism advocated in the metaphysical context of the *Meditations* (at least up to Meditation 5: in Meditation 6 Descartes tries to mitigate his radical dualism), and is much closer to the naturalistic account of cognitive states offered in the *Rules for the Direction of the Native Intelligence* and the *Treatise on Man*.

Chronology

1626–8   Resumes work on the *Rules*, the focus now being on questions
          of the mechanistic construal of perceptual cognition, and the
          problem of legitimating mathematical operations. He finally
          abandons the *Rules* in 1628. At the end of 1628, he settles in the
          Netherlands

1629–0   Begins work on a number of metaphysical questions, as well as
          devising a machine for grinding hyperbolic lenses. From August
          1629 onwards, other projects are gradually abandoned as he
          tries to explain the meteorological phenomenon of parhelia,
          which by the end of 1629 has grown into an attempt to account
          for 'the whole physical world'

1630–2   The *Dioptrics* and the *Meteors* are completed in draft. While in
          Amsterdam he visits butchers' shops daily to retrieve pieces for
          dissection. In May 1632 he moves to Deventer, partly to avoid
          interruptions to his work, as he works intensely on physical
          optics, the laws of motion, and the outlines of a cosmology. The
          unfinished draft of the *Treatise on Light* dates from 1632

1632–3   Turns to the *Treatise on Man*, setting out a mechanistic physi-
          ology and a theory of the body as an automaton. Between July
          and November 1633, he prepares his treatises for publication,
          only to hear in November of the Inquisition's condemnation of
          Galileo, at which point, in obvious despair, he abandons plans
          to publish

1634–6   He prepares final drafts of the *Dioptrics* and the *Meteors*, and
          starts to work on a preface to them, which will become the
          Discourse on Method: the *Geometry*, which will accompany these
          texts, is put together from earlier drafts while the other treatises
          are being printed

1637–9   The *Discourse* and the three *Essays* are published in June 1637

1639–40   Works on the *Meditations*, returning to Leiden in April 1640 to
          supervise a preliminary printing of the *Meditations*

1641–3   The *Meditations* are published in 1641, together with six sets of
          objections and replies. After giving up the idea of writing a
          dialogue (*The Search for Truth*), he begins work on a compre-
          hensive exposition of his philosophy in textbook form, the
          Principles of Philosophy, at the end of 1641. The second edition
          of the *Meditations*, with a seventh set of objections and replies
          and a letter to Dinet, in which Descartes defends himself

against attacks on the orthodoxy of the *Meditations*, appears in 1642. In response to Descartes' long attack on him in the *Letter to Voetius*, published in May 1643, Voetius succeeds in having the Council of Utrecht summon Descartes, and he is threatened with expulsion and the public burning of his books. He seeks refuge in the Hague

1643–6   Starts an affectionate and fruitful correspondence with Princess Elizabeth of Bohemia, focusing on his account of the passions. The *Principles*, four parts of its originally projected six complete, is published by Elzevier in the middle of 1644. A good deal of his time is taken up with dissection of animals and studying the medicinal properties of plants. By 1646 he has a draft of the *Passions of the Soul*

1647–8   Is condemned by Revius and other theologians at the University of Leiden in early 1647. French translations of the *Meditations* and the *Principles* are published in 1647, and his attack on his erstwhile disciple Regius, the *Notes on a Certain Programme*, appears at the beginning of 1648. In 1647–8 he works on the unfinished *Description of the Human Body*

1649–50   Leaves for the court of Queen Christiana of Sweden on 31 August 1649. The *Passions of the Soul* appears in November 1649. He dies of pneumonia in Stockholm, on 11 February 1650

Further reading

Annie Bitbol-Hespériès and Jean-Pierre Verdet, *René Descartes: Le Monde, L'Homme* (Paris, 1996) provides the French text of the *Treatise on Light* and the *Treatise on Man*, with an introduction and annotations. Among recent works in English that deal in detail with both the works in the context in which they were written are Stephen Gaukroger, *Descartes: An Intellectual Biography* (Oxford, 1995), and William R. Shea, *The Magic of Numbers and Motion* (Canton, Mass., 1991). On the politico-religious background to the abandonment of *Le Monde*, the following is useful: Richard J. Blackwell, *Galileo, Bellarmine, and the Bible* (Notre Dame, 1991).

Among some of the more useful general tools are John Cottingham, *A Descartes Dictionary* (Oxford, 1993), and the much more specialised work by Etienne Gilson, *Index scolastico-cartésien* (2nd edn. Paris, 1979). The best bibliographical source is Gregor Sebba, *Bibliographia Cartesiana: A Critical Guide to the Descartes Literature, 1800–1960* (The Hague, 1964): material published since 1960 has not received anything like the careful treatment offered by Sebba.

On Descartes' work in physical theory generally, see: Gerd Buchdahl, *Metaphysics and the Philosophy of Science* (Oxford, 1969); Desmond Clarke, *Descartes' Philosophy of Science* (Manchester, 1982); Pierre Costabel, *Démarches originales de Descartes savant* (Paris, 1982); Peter Damerow, Gideon Freudenthal, Peter McLaughlin, and Jürgen Renn, *Exploring the Limits of Preclassical Mechanics* (New York, 1992); Alan Gabbey, 'Force and Inertia in the Seventeenth Century: Descartes and Newton', in Stephen Gaukroger (ed.), *Descartes: Philosophy, Mathematics and Physics* (Brighton, 1980), 230–320; Daniel Garber, *Descartes' Metaphysical Physics* (Chicago, 1992); Emily Grosholz, *Cartesian Method and the Problem of Reduction*

(Oxford, 1991); Martial Gueroult, 'The Metaphysics and Physics of Force in Descartes', in Stephen Gaukroger (ed.), *Descartes: Philosophy, Mathematics and Physics* (Brighton, 1980), 196–229; John Schuster, 'Descartes and the Scientific Revolution, 1618–1634', Princeton University Ph.D. thesis, University Microfilms reprint (2 vols., Ann Arbor, [1977]); J. F. Scott, *The Scientific Work of René Descartes* (London, 1952). Specifically on Descartes' optics, the following are especially important: I. A. Sabra, *Theories of Light from Descartes to Newton* (Cambridge, 1981); A. Mark Smith, 'Descartes' Theory of Light and Refraction: A Discourse on Method', *Transactions of the American Philosophical Society* vol. 77, part 3 (1987), 1–92. On the more specific questions of the rainbow and the propagation of light see respectively Charles B. Boyer, *The Rainbow* (Princeton, 1987), and Alan E. Shapiro, 'Light, Pressure, and Rectilinear Propagation: Descartes' Celestial Optics and Newton's Hydrostatics', *Studies in History and Philosophy of Science* vol. 5 (1974), 239–96. On the question of visual cognition, and the problematic ch. 1 of the *Treatise on Light*, a good starting-point is John Yolton, *Perceptual Acquaintance* (Oxford, 1984). Specifically on Descartes' cosmology, see Eric Aiton, *The Vortex Theory of Planetary Motions* (London, 1972).

On Cartesian physiology generally, the most comprehensive single account is Annie Bitbol-Hespériès, *Le principe de vie chez Descartes* (Paris, 1990), but Thomas Steele Hall's *René Descartes: Treatise of Man* (Cambridge, Mass., 1972) offers sufficiently comprehensive and broad-ranging annotations to give a good overall picture. Similarly, his *Ideas of Life and Matter* (2 vols., Chicago, 1969) gives a good, if basic, account of the background to the development of physiology. On the relation between psychology and physiology, see Gary Hatfield, 'Descartes' Physiology and its Relation to his Psychology', in John Cottingham (ed.), *The Cambridge Companion to Descartes* (Cambridge, 1992), 335–70; G. Rodis-Lewis, *L'anthropologie cartésienne* (Paris, 1990); Amélie Oksenberg Rorty, 'Descartes on Thinking with the Body', in John Cottingham (ed.), *The Cambridge Companion to Descartes* (Cambridge, 1992), 371–92. On the question of Descartes' discovery of reflex action, a dissenting view is to be found in Georges Canguilhem, *La formation du concept du réflex aux XVIIe et XVIIIe siècles* (Paris, 1955). On the question of animals as machines see Leonora Cohen Rosenfield, *From Beast-Machine to Man-Machine*, revised edn (New York, 1968). On Descartes' account of memory, see John Sutton, *Philosophy and Memory Traces* (Cambridge, 1998), which also contains material on Descartes' account of animal spirits: on the background to this latter question, see Daniel P. Walker, 'Medical Spirits in Philosophy and Theology from Ficino to Newton', in *Arts du spectacle et histoire des idées. Recueil offert en hommage à Jean Jacquot* (Tours, 1984), 287–300.

Note on the texts

The *Treatise on Light* was first published as *Le Monde de Mr. Descartes ou le Traité de la Lumière* in Paris in 1664, and the *Treatise on Man* first appeared two years earlier in Leyden, in Latin translation by Florentino Schuyl, as *Renatus Descartes de Homine*. In 1664 Clerselier, who had access to the original manuscripts, brought out an edition of *L'Homme*, and in 1677 an edition of both texts. This has been the basis for modern editions, the standard modern edition being that of Charles Adam and Paul Tannery, *Oeuvres de Descartes* (new edn, Paris, 1974–86). The text as given in volume 10 of this edition (abbreviated to AT) is the one I have used for the translations given here, and references to the page numbers in the AT edition are given in the margins.

The illustrations are not Descartes' own, although those in the *Treatise on Light* are undoubtedly based on sketches, no longer extant, by Descartes. Those in the *Treatise on Man* derive from the treatise's first editor, Schuyl. Clerselier commissioned his own illustrations, which I have reproduced here, and these are slightly different from those of the first editions.

As for the supplementary material, the *Dioptrics* and the *Meteors* were published in Descartes' lifetime, and the texts I have used are from volume 6 of AT. The *Description of the Human Body* first appeared as part of Schuyl's edition of the *Treatise on Man*, and I have used the text as given in AT volume 10.

The *Treatise on Light* and related material

Treatise on Light and other principal objects 3
of the senses

Chapter 1

On the difference between our sensations[1] and the things that produce them[2]

In putting forward an account of light, the first thing that I want to draw
to your attention is that it is possible for there to be a difference between
the sensation that we have of it, that is, the idea that we form of it in our
imagination through the intermediary of our eyes, and what it is in the
objects that produces the sensation in us, that is, what it is in the flame or
in the Sun that we term 'light'. For although everyone is commonly convinced that the ideas that we have in our thought are completely like the
objects from which they proceed, I know of no compelling argument for
this. Quite the contrary, I know of many observations which cast doubt 4
upon it.

As you know, the fact that words bear no resemblance to the things they
signify does not prevent them from causing us to conceive of those things,

---

[1] I have translated the term *sentiment* by 'sensation'. Although Descartes will include pains among
our sensations in the *Treatise on Man*, the qualification that a sensation is formed 'through the
intermediary of our eyes' restricts sensations to ideas caused by external objects. However, sensation should not be taken in the sense of mere sensation, as opposed to perception, something which
involves judgement, for *sentiments* can involve judgement, and indeed typically involve judgements
in the case of human sensations. The sensations of automata do not involve judgement, and cases
of human sensation in which there is no attentiveness, such as our perception of objects at the
extremes of our visual field, seem to be treated on a par with an automaton's sensation (see AT i.
413; CSM iii. 61–2).

[2] The chapter headings, and possibly even the division into chapters, were probably the work of
Clerselier. I give the chapter headings of the 1677 edition; the 1664 chapter headings, which are
probably the work of an early copyist, are given in the notes where these differ.

often without our paying attention to the sounds of the words or to their syllables. Thus it can turn out that, having heard something and understood its meaning perfectly well, we might not be able to say in what language it was uttered. Now if words, which signify something only through human convention, are sufficient to make us think of things to which they bear no resemblance, why could not Nature also have established some sign which would make us have a sensation of light, even if that sign had in it nothing that resembled this sensation? And is it not thus that Nature has established laughter and tears, to make us read joy and sorrow on the face of men?[3]

But perhaps you will say that our ears really only cause in us sensory awareness of the sound of the words, and our eyes only sensory awareness of the countenance of the person laughing or crying, and that it is our mind which, having remembered what those words and that countenance signify, represents this to us at the same time. I could reply to this that, by the same token, it is our mind that represents to us the idea of light each time the action that signifies it touches our eye. But rather than waste time arguing, it is better to give another example.

Do you think that, when we attend solely to the sound of words with-

---

[3] This is a key passage, but it is too compact for us to say with certainty exactly what Descartes has in mind. In discussing perceptual cognition in earlier works such as the *Rules*, Descartes focused on the 'perceptual' side of the question, whereas here he clearly wants to say something about the 'cognition' side. The former he construes in terms of mechanical-physiological process, as is clear from the *Treatise on Man*. Here he construes the latter in linguistic terms, so that visual cognition – knowing something by virtue of seeing it – is considered not in terms of seeing and understanding a picture but in terms of hearing and understanding a word or a sentence: any element of resemblance between the thing perceived and our cognitive representation of the thing is completely purged. What happens when we understand what another person says is that the idea in that person's mind is conveyed to our mind: the idea or thought is encoded in language and then decoded by our mind. The words that encode the idea clearly do not resemble it, but they just as clearly do represent it. So far so good, but once we apply this model to the visual perception of objects we immediately face a disanalogy. For in what sense is there an idea conveyed to our mind when we see something? Are there ideas in nature, which nature itself encodes, or which God has encoded there? We can think of the question in terms of Descartes' terminology of signs. For Descartes, language consists of conventional signs; these signs signify thoughts or ideas for the purpose of conveying those thoughts or ideas to another person who understands the signs. In the case of visual perception, what are the analogues of the speaker's thoughts, the conventional linguistic signs, and the hearer's thoughts? One might be tempted to say that they are, respectively, natural objects, the natural signs by which information about these natural objects is conveyed to us visually (namely light), and the perceiver's thoughts. But this is not consistent with the way in which Descartes construes what happens. He tells us that there is in nature a sign which is responsible for our sensation of light, but which is not itself light, and which does not resemble light: all there is in nature is motion. Motion is the sign, and what is signified is what is experienced in the perception, namely light. This makes it look as if what is signified in nature is something that exists only in our mind, a view we could hardly ascribe to Descartes.

out attending to their signification, the idea of that sound which is formed in our thought is at all like the object that is the cause of it? A man opens his mouth, moves his tongue, and breathes out: I see nothing in all these actions which is in any way similar to the idea of the sound that they cause us to imagine. And most philosophers maintain that sound is only a certain vibration of the air striking our ears.[4] Thus if the sense of hearing transmitted to our thought the true image of its object, then instead of making us think of the sound, it would have to make us think about the motion of the parts of the air that are vibrating against our ears. But as not everyone will, perhaps, wish to follow what the Philosophers[5] say, so I shall offer another example.

Of all our senses, touch is the one considered least deceptive and the most secure; so if I show you that even touch leads us to conceive many ideas which do not resemble in any way the objects that produce them, I believe you should not find it strange when I say that the same holds for sight. Now everyone knows that the ideas of tickling and pain which are formed in our thought when bodies from outside touch us bear no resem- 6 blance at all to these. One passes a feather lightly over the lips of a child who is falling asleep and he feels himself being tickled: do you think that the idea of tickling which he conceives resembles something in the feather? A soldier returns from battle. During the heat of the combat he could have been wounded without being aware of it. But now, as he begins to cool down he feels pain and believes that he has been wounded: a surgeon is called and examines him once his armour has been removed; in the end, it is discovered that what he was feeling was just a buckle or strap which, being caught under his armour, was pressing on him and causing his discomfort. If his sense of touch, in causing him to feel this strap, had impressed its image in his thought, there would not have been any need for the surgeon to show him what he was feeling.

Now I can see nothing which compels us to believe that what it is in objects that gives rise to the sensation of light is any more like that

[4] An early version of the vibration theory had been held by the Coimbra commentators. See the texts given in Gilson, *Index scolastico-cartésien* (2nd edn, Paris, 1979), nos. 424 and 425. A related 'corpuscular' theory of sound had been developed by Descartes' early mentor Isaac Beeckman in the second decade of the seventeenth century, and Mersenne developed this approach in detail in the 1620s and 1630s. Here was a rare case of relatively common ground in natural philosophy.
[5] The phrase 'les Philosophes' usually refers specifically to scholastic philosophers, and as often as not to the late scholastic Jesuit philosophers – Suárez, Toletus, Fonseca, and the Coimbra commentators – from whose commentaries Descartes had learned his philosophy at La Flèche.

sensation than the actions of a feather or a strap are like a tickling sensation and pain. Nevertheless, I have not adduced these examples to convince you absolutely that light is something different in objects from what it is in our eyes, but only to raise a doubt about it for you, to prevent you being biased in favour of the contrary view, so that we can examine together what light is.

7

Chapter 2

What the heat and the light of fire consist in[6]

I know of only two kinds of bodies in the world in which light is found, namely the stars, and flame or fire.[7] And because there is no doubt that stars are further from human knowledge than fire or flame, I shall first try to explain what I notice with respect to flame.

When it burns wood or other similar material we can see with our eyes[8] that it moves the small parts of the wood, separating them from one another, thereby transforming the finer parts into fire, air, and smoke, and leaving the larger parts as ashes. Someone else may if he wishes imagine the 'form' of fire, the 'quality' of heat, and the 'action' of burning to be very different things in the wood.[9] For my own part, I am afraid of going astray if I suppose there to be in the wood anything more than what I see must necessarily be there, so I am satisfied to confine myself to conceiving the motion of its parts. For you can posit 'fire' and 'heat' in the wood and make it burn as much as you please: but if you do not suppose in addition that some of its parts move or are detached from their neighbours then I cannot imagine that it would undergo any alteration or change.[10] On the other hand, take away the 'fire', the 'heat', and keep

8 the wood from 'burning'; then, provided only that you grant me that

---

[6] The heading in the 1664 edition is: *What it is in fire that burns, heats, and illuminates.*

[7] The obvious omission here is phosphorescent phenomena.

[8] That is, presumably, without the help of a magnifying glass. The phenomenon is macroscopic, even though it turns out that it must be explained in micro-corpuscularian terms.

[9] Descartes is referring here to the Aristotelian account of fire. Aristotle treats fire as one of the four elements in Book II of *De Generatione et Corruptione*, that element characterised by the qualities hot and dry. The elements can be transformed into one another by a change in their qualities, and he gives the example of fire and water being transformed into air and earth. The (qualitatively characterised) type of change involved in the transformation is the main subject of Aristotle's discussion. Nevertheless, it is not Aristotle's own account that Descartes has principally in mind here but that of the late scholastic commentators. Gilson traces reasonably direct sources in Suárez and Eustache de Saint Paul in his *Index*, nos. 211 and 392.

there is some power that violently removes its more subtle parts and separates them from the grosser parts, I consider that this alone will be able to bring about all those changes that we observe when the wood burns.

Now since it does not seem possible to conceive of a body moving another unless it itself is moving, I conclude from this that the body of the flame which acts against the wood consists of minute parts, which move independently of one another with a very quick and violent motion; and as they move in this way, they push against and move those parts of the body that they touch and which do not offer them too much resistance. I say that its parts move independently of one another because although often many of them act together to bring about a single effect, we see nonetheless that each of them acts on its own against the bodies they touch. I say also that their motion is very quick and very violent, for being so minute that we cannot distinguish them by sight, they would not have the force to act against other bodies if the quickness of their motion did not compensate for their lack of size.[11]

I add nothing about the direction in which each moves. For when you consider that the power to move and the power that determines in what direction the motion must take place are two completely different things, 9 and can exist one without the other (as I have explained in my *Dioptrics*[12]), then you will have no difficulty recognising that each part moves in the manner made least difficult for it by the disposition of the bodies surrounding it.[13] And in one and the same flame, there can be some parts going up, and others down, some in straight lines, some in circles; they can move in every direction without altering its nature at all. Thus if you see almost all the parts tending upwards, you need not think

[10] Aristotle had maintained that local motion is involved in every other kind of change in his *Physics* (208ª32 and 260ᵇ22). Descartes now moves from this relatively uncontentious claim to something more like the view that the other forms of change are reducible to local motion, something which Aristotle and the scholastic tradition completely reject.
[11] How the quickness of their motion can 'compensate' for their small size is not set out in the text. The simplest relation suggested by what Descartes says is that the force involved is to be measured by size × speed, but Descartes thinks of force in so many different ways, and is normally so reluctant to consider speeds, that it is not possible to say just what the relationship here is.
[12] See translation of Discourse 2 of the *Dioptrics*, below.
[13] The implicit principle that the part of the flame will always take the path which offers least resistance is problematic. On a literal reading of this principle, light (which will be treated on a par with fire) transmitted through air would always be reflected when it met an opaque surface, for the opaque surface would always resist its motion more than the air. This alone would rule out a literal reading. What the intended reading of 'least resistance' is in the present context is obscure.

7

that this is for any reason other than that the bodies touching them are almost always disposed to offer them greater resistance in any other direction.[14]

Once we appreciate that the parts of the flame move in this way, and that to understand how the flame has the power to consume the wood and to burn it, it is enough to conceive of their motions, I ask you to consider whether this is not also sufficient for us to understand how the flame provides us with heat and light.[15] For if this is the case, the flame will need possess no other quality, and we shall be able to say that it is this motion alone that is called now 'heat' and now 'light', according to the different effects it produces.

As regards heat, it seems to me that our sensation of it can be taken as a kind of pain when it is violent, and sometimes as a kind of tickling, when it is moderate.[16] Since we have already said that there is nothing outside our thought which is similar to the ideas which we conceive of tickling and pain,[17] we can well believe that there is nothing that is similar to that which we conceive of as heat; rather, anything that can move the minute parts of our hands or of any other place in our body can arouse this sensation in us. There are many observations which support this view. For merely by rubbing our hands together we can heat them, and any other body can also be heated without being placed close to a fire, provided only that it is shaken and rubbed in such a way that many of its minute parts are moved and thereby can move the minute parts of our hands.

As regards light, it can also be conceived that this same motion in the flame suffices to make us sense it. But since the main part of my project is to deal with this, I want to try to explain it at length when I resume discussion of this matter.

---

[14] The relevant contrast here is with Aristotle's theory, whereby flames move upwards because the natural place of fire is upwards. See, for example, *De Caelo* 311ª15ff.

[15] The cases of motion producing combustion and motion producing heat and light are, nevertheless, very different. As is evident from the next paragraph, there is a difference of kind between the motion that produces heat and our sensation of heat, but there is no such difference in the case of combustion.

[16] A mechanistic account of pain and tickling will be provided in the *Treatise on Man*, AT xi. 143–4, p. 119 below.

[17] It is tempting to translate *concevoir* here as 'have', and to speak simply of the idea we have of tickling, rather than the idea we conceive of tickling, but 'have' does not convey the active ingredient in conceiving an idea, which is important in Descartes' account.

Chapter 3

Hardness and fluidity[18]

I believe that there are innumerable different motions which endure perpetually in the world. After having noted the greatest of these – those which bring about the days, months, and years – I take note that the terrestrial vapours unceasingly rise to and descend from the clouds, that the air is forever agitated by the winds, that the sea is never at rest, that 11 springs and rivers flow ceaselessly, that the strongest buildings eventually fall into decay, that plants and animals are always either growing or decaying: in short, that there is nothing anywhere which is not changing. From this it is evident to me that the flame is not alone in having many minute parts in ceaseless motion, but that every other body has such parts, even though their actions are not as violent and, because of their small size, they cannot be perceived by any of our senses.

I do not pause to seek the cause of their motions, for it is enough for me to take it that they began to move as soon as the world began to exist. And that being the case, I reason that their motions cannot possibly ever cease, or even change in any way except in respect of their subject. That is to say, the strength or power found in one body to move itself may pass wholly or partially to another body and thus no longer be present in the first, but it cannot entirely cease to exist in the world.[19] My arguments had satisfied me on this point, but I have not yet had the opportunity to present them to you. In the meantime you might care to imagine, along with most of the learned,[20] that there is some prime mover which, rolling 12 around the world at an incomprehensible speed,[21] is the origin and source of all the other motions found therein.

[18] The heading in the 1664 edition is: *Where the variety, duration and cause of motion are examined, with the explication of the hardness and fluidity of bodies in which these are found.*

[19] For Aristotle, new motions can come into existence, and motion can be dissipated out of existence. Descartes here denies this, albeit by fiat, effectively stating a conservation law. We must be careful about what exactly is conserved, however. It would seem to be not so much the total quantity of motion as the total quantity of the strength [*vertu*] or power [*puissance*] by which a body moves, or, in more convenient terminology, the total quantity of the force of motion. In virtue of conservation of the total quantity of force of motion there will be conservation of the total quantity of motion, but the two must be distinguished, partly because the relations between motion and force of motion in Descartes' natural philosophy are complex, and partly because it is important to realise that conservation of motion is due to conservation of force of motion when one comes to assess the relation between kinematic and dynamic considerations in Descartes. His statement of conservation here involves forces, and so is dynamic rather than kinematic.

[20] The term Descartes uses here is 'Doctes', indicating above all scholastic thinkers.

[21] Gilson gives sources for this doctrine in the Coimbra commentators: see Gilson, *Index*, no. 308.

Now this consideration leads to a way of explaining all the changes that occur in the world, and all the variety that appears on the earth; but I shall confine myself here to speaking of those that bear on my topic.

The first thing I want to call to your attention is the difference between bodies that are hard and those that are fluid. To this end, consider that every body can be divided into extremely small parts. I am not interested in deciding whether the number of these is infinite or not; at least with respect to our knowledge, it is certain that it is indefinite and that we can suppose that there are several million of them in the smallest grain of sand visible to the eye.

And note that if two of these minute parts are touching one another and are not in the process of moving away from each other, then a force, no matter how small, is needed to separate them; for once they are so positioned, they would never be inclined to dispose themselves differently. Note also that twice as much force is needed to separate two of them than is needed for one, and a thousand times as much to separate a thousand of them. Consequently, if one had to separate several million of them at once, as is perhaps necessary in breaking a single hair, it is not surprising that a significant force is required.[22]

By contrast, if two or more of these minute parts only touch in passing and while they are in the process of moving one in one direction and one in the other, it is certain that it will require less force to separate them than if they were completely stationary, and indeed none at all if the motion with which they are able to separate themselves is equal to or greater than that with which one wishes to separate them.

Now I detect no difference at all between hard bodies and fluid bodies except that the parts of the one can be separated from the whole much more easily than those of the other. Thus, to make the hardest body imaginable, I think it would be enough for all the parts to touch each other, with no space remaining between any two and none of them in the process of moving. For what glue or cement can one imagine beyond this with which to hold the one to the other?

Moreover, I think that it is enough, to make the most fluid body

[22] One should not imagine something like a chain of a hundred links each of which can bear exactly ten pounds here, for if eleven pounds is enough to break any of the links it will not matter how many other links it is attached to: the chain will not support the weight. Rather, one must think of each of the links, not as being attached to one another, but being each attached directly to the weight. In this case the weight is evenly distributed throughout the links, and such links will bear (roughly) a hundred times the weight one will bear.

imaginable, that all its most minute parts be moving away from one another in the most diverse ways and as quickly as possible, even though in that state they are quite able to touch one another on all sides, and to arrange themselves in a space as small as if they were motionless. Finally, 14 I believe that every body approaches these two extremes to a greater or lesser degree, depending on the degree to which its parts are in the process of separating themselves from one another. And this judgement is corroborated by everything I have cast my eye on.

Flame, whose parts – as I have already said – are perpetually agitated, is not only fluid, but renders most other bodies fluid. And note that when it melts metals, it acts with a power no different from that by which it burns wood.[23] But because the parts of the metal are all approximately equal [in size], it cannot move one without the other, and consequently it forms completely fluid bodies from them. The parts of wood, by contrast, are unequal in such a way that the flame can separate out the smaller of them and make them fluid – that is, it can cause them to fly away as smoke – without thereby agitating the larger parts.

After flame, there is nothing more fluid than air, and one can see with the naked eye that the parts move separately from one another. For if you take the trouble to watch those minute bodies that are commonly called atoms which appear in rays of sunlight, you will see that, even when there is no wind stirring them up, they flutter about incessantly in a thousand different ways.[24] The same kind of thing can also be experienced in all the grosser liquids if differently coloured ones are mixed together in order 15 that their motions might be distinguished more easily. And finally this can be experienced very clearly in acids, [25] when they move and separate the parts of some metal.

---

[23] The task that Descartes has set himself here is, with hindsight, an impossible one. His aim is to account for the traditional four elements – earth, air, fire, and water – as the four states of a single substance. Earth, water, and air can be taken as solid, liquid, and gaseous states respectively, and there are clearly prospects for success in treating these as different states of the one substance. But fire cannot be fitted into this schema, and his attempt to draw parallels between the liquefaction of solids and the combustion of solids, although ingenious, is doomed, and never rises above the level of the speculative.

[24] The 'atoms' that Descartes refers to here are of course dust particles which, in common with many of his contemporaries, he takes to be minute particles of air.

[25] Descartes' term '*les eaux fortes*' has a rather broad variety of meanings. Most literally it is a translation of the Renaissance Latin term for nitric acid, *aqua fortis*, but virtually any liquid which had, or was thought to have, the power of dissolving substances could come under the term, and sixteenth- and seventeenth-century alchemists regularly treated mercury as the basic *eau forte*. Nevertheless, nitric acid is the most likely contender here as it was widely available owing to its use in etching copper plates.

But at this point you may ask, if it is solely the motion of the parts of the flame that cause it to burn and make it fluid, why the motion of the parts of air, which also make it extremely fluid, give it no power at all to burn but, quite the contrary, make it such that our hands can hardly feel it? To this I reply that one must take account not only of the speed of motion, but also the size of the parts. It is the smaller ones that make the more fluid bodies, but it is the larger ones that have more force to burn and, in general, to act on other bodies.

Note, by the way, that here, and always from here onwards, I shall take a single part to be everything that is joined together and which is not in the process of separating, even though the smallest parts could be divided easily into many smaller ones; thus a grain of sand, a stone, a rock, indeed the whole earth itself, can from here on be taken as a single part, in so far as we are considering here only a completely simple and completely equal motion.

16 Now if, among the parts of the air, there are some which are very large in comparison with others, as are the atoms that are seen there, they also move very slowly; and, if there are some that move more quickly, they are also the smallest. But if, among the parts of the flame, there are some that are smaller than those in air, there are also larger ones, or at least there is a larger number of parts of the same size as the largest parts of air, and they move much more quickly. Consequently these alone have the power to burn.

That there are smaller parts may be conjectured from the fact that many bodies that they penetrate have pores so narrow that even air cannot enter them. That there are larger parts, or parts as large but in greater number, is seen clearly from the fact that air alone is not enough to keep the flame burning.[26] That they move more quickly is sufficiently evident from the violence of their action. And finally, that it is the largest of these parts that have the power to burn, and not the others, is apparent from the fact that the flame that issues from brandy, or from other very subtle bodies, hardly burns at all, while that which comes from hard and heavy bodies is very hot.[27]

---

[26] Literally, air alone is not enough to 'nourish' the flame. The connection between air's inability to keep a flame alight and the claim that its largest parts must be larger than, or more numerous than, those of air is obscure. The metaphor of nourishment seems to be the key to what Descartes has in mind here: we can only nourish ourselves by breaking down relatively large things. The ability of something to nourish seems to be associated with its amenability to being broken down into smaller parts.

[27] Descartes attempts to spell out the structural differences between various kinds of body in the *Meteorology*: see AT vi. 233–4. This material may date from as early as the time of composition of the present discussion.

Chapter 4

On the void, and how it comes about that our senses do not perceive certain bodies[28]

But we need to examine in greater detail why, although it is as much a body as any other, air cannot be sensed as easily as other bodies; and in doing this we shall free ourselves from an error which has been a prejudice since childhood, when we believed that the only bodies around us were those that we could perceive, and consequently that, if air were one of these then, because we perceive it so faintly, it must at least not be as material and solid as those we sense more clearly.

On this topic, the first thing I would like you to note is that all bodies, whether hard or fluid, are made from the same matter, and that it's impossible to conceive of the parts of this matter ever composing a more solid body, or occupying less space, than they do when each of them is touched on all sides by the others surrounding it. From this it seems to me to follow that if there could be a void anywhere it must be in hard bodies rather than fluid ones;[29] for it is obviously much easier for the parts of the latter to press and arrange themselves against one another, because they are moving, than it is for those of the former, which are motionless.

When you put powder in a jar, for example, you shake and pound it to make room for more powder; but if you pour liquid into it, it immediately arranges itself in the smallest space into which one can put it.[30] And indeed, if you think in this respect of some of the experiments that philosophers commonly use to show that there can be no void in nature, you will readily appreciate that all those spaces that people consider

---

[28] The heading in the 1664 edition is: *What judgement we must make about the void, and the reason why our senses are not aware of certain bodies.*

[29] The traditional Atomist explanation of differences in density was in terms of the amount of void that existed between the atoms making up a body. The more empty space between atoms, the less the body's density. In undermining the Atomist view, Descartes shifts the focus to fluidity. We naturally think in terms of an inverse proportion between density and fluidity, and the present argument is designed to show that the behaviour of hard bodies is incompatible with the postulation of a void.

[30] There are two problems with this counter-example to the traditional Atomist account. First, given that Descartes has made the ease with which something can be penetrated or cut a criterion for its being a fluid, powders should be treated as fluids, not as solids. Secondly, if one does treat powders as a kind of solid, as Atomists presumably do, then one has to distinguish between the behaviour of powders, which can be pounded into a smaller volume, and rigid solids, which cannot be.

empty, and where we perceive only air, are no less full – and of the same matter – as the spaces where we perceive other bodies.

For pray tell me why on the one hand Nature would cause the heaviest bodies to rise and the most solid to break, as we can see it doing in certain machines, rather than to allow their parts to cease to touch one another or to touch other bodies, and yet on the other allow the parts of air, which are easy to bend and arrange in every way, to remain next to each other without being touched on each side, or without there being any body between them which they touch. Could one really believe that, on the one hand, the water in a well has to rise, contrary to its natural inclination, merely in order that the pipe of a pump may be filled, and that, on the other hand, the water in the clouds does not have to fall in order to fill the spaces here below, if there were even the least void between the parts of the bodies they contain?[31]

But you could bring up a more considerable difficulty here, namely, that it does not seem that the parts composing liquid bodies can move 19 about incessantly as I have said they do, unless there is some empty space between them, at least in the places they vacate as they move about. I would have trouble replying to this had I not learned from a variety of observations that all motions that occur in the world are in some way circular. That is, when a body leaves its place, it always enters into that of another, and this latter into that of another, and so on to the last body, which at the same instant occupies the first. Thus there is no more of a void between bodies when they are moving than when they are at rest. And note here that for this to happen it is not necessary that all the parts of the bodies that move together be arranged exactly in a ring, as in a true circle,[32] or even that they be of equal size and shape, for any such inequalities can easily be compensated for by other inequalities in their speeds.

We do not usually notice these circular motions when bodies are moving in air because we are accustomed to thinking of air as being just empty space. But look at fish swimming in the pool of a fountain: if they do not come too near the surface of the water, they cause no motion at all in it, even though they are passing beneath it at great speed. It is clearly apparent

---

[31] In other words, a pump can raise water against its natural inclination (its weight tending to cause it to fall, not rise) because a vacuum would be formed unless the water rose: and this being the case, if there were empty spaces between the parts of matter on the earth, we would have even more reason to expect that they would draw water out of the clouds, since this would be in keeping with the tendency of water to fall, due to its weight.

[32] In fact, all that is required is that the motions form a closed curve.

from this that the water they push before them does not push all the water in the pool indiscriminately, but only that which can best serve to perfect 20 the circle of the fishes' motion and to occupy the place they vacate.

And this observation is sufficient to show the ease and familiarity of such circular motions to Nature. But I now want to put forward another observation, which shows that no motion ever occurs which is not circular. When the wine in a cask does not flow through an opening at the bottom because the top is shut tight, it is improper to say, as is commonly done, that this takes place because of 'fear of a void'. It is well known that the wine has no mind with which to fear anything, and even if it did, I do not know why it should fear a void, which is wholly chimeral. Instead, what we must say is that the wine cannot leave the cask because outside everything is completely full, and the part of the air whose place the wine would occupy if it were to flow out can find nowhere else in the universe to occupy, unless an opening is made in the top of the cask through which the air can rise in a circle into its place.

Nevertheless, I do not want to say categorically that there is no void in Nature. I fear that my treatise would be too lengthy if I were to undertake to explain the matter at length, and the observations of which I have spoken are not sufficient to secure it, although they are enough to persuade us that those spaces where we sense nothing are filled with the same matter, 21 and contain at least as much of that matter, as those occupied by the bodies that we perceive. Thus, when a vessel is full of gold or lead, for example, it contains no more matter than when we think it empty. This may seem strange to those people whose reasoning extends no further than their fingertips, and who think there is nothing in the world other than what they touch. But once you have given a little consideration to what makes us perceive or not perceive a body with our senses, I am sure that you will find that there is nothing incredible in this. For you will recognise clearly that, far from all the things around us being perceivable, on the contrary it is those that are there most of the time that can be perceived the least, and those that are there all of the time can never be perceived at all.

The heat of our heart is very great, but we do not feel it because it is always there. The weight of our body is great, but it does not discomfort us. We do not even feel the weight of our clothes, because we are used to wearing them. The reason for this is clear enough: we cannot perceive a body by our senses unless it is the cause of some change in our sense organs – that is, unless it moves in some way the small parts of matter of 22

15

which those organs are composed. The objects that are not always pre-sent can do this well enough, provided they have enough force; for if they damage something in the sense organs while acting upon them, that can be repaired afterwards by nature, when they are no longer acting. But as for those objects which continually touch us, even if they had the power to induce a change in our senses and to move some parts of their matter, they would have to have moved them and separated them completely from the others at the beginning of our life, and in this way they would have left there only the parts that completely resist their action, and with-out which they could not be perceived by our senses in any way. You can see from this that it is no wonder that there are many spaces around us in which we do not perceive any body by our senses, even though they con-tain bodies no less than the spaces in which we perceive them the most.

But it must not be thought that the gross air that we draw into our lungs while breathing – the air which turns into wind when set in motion, which seems hard when enclosed in a balloon, and which is composed only of exhalations and smoke – is as solid as water or earth. Here we must follow the common opinion of the Philosophers, who all maintain that it is rarer, and we can tell this easily from experience. For when the parts of a drop of water are separated from one another by the agitation of heat,
23 they can make up much more of this air than could be contained in the space that held the water. From this it follows with certainty that there are many small gaps between the parts of which the air is composed; for there is no other way to conceive a rare body. But because these gaps cannot be empty, as I said above, I conclude from this that there must be other bodies, one or many, mixed with the air, and these bodies fill the tiny gaps left between the parts as tightly as possible. It only now remains for me to consider what these other bodies can be, and after this I hope it will not be difficult to understand what may be the nature of light.

Chapter 5

On the number of elements and their qualities[33]

The Philosophers maintain that above the clouds there is a kind of air much subtler than ours, which is not composed of terrestrial vapours, as

---

[33] The heading in the 1664 edition is: *The reduction of the four Elements to three, with an explanation and establishment of them.*

our air is, but constitutes an element in itself. They say too that above this air there is yet another body, more subtle still, which they call the element of fire. And they add that these two elements are mixed with water and earth to make up all the bodies below.[34] Thus I shall merely be following their opinion if I say that this subtler air and this element of fire fill the gaps between the parts of the gross air that we breathe, so that these bodies, interlaced with one another, make up a mass as solid as any body can be. 24

But so that you might understand my thought on this subject better, and not think that I am forcing you to believe everything the Philosophers tell us about the elements, I must describe them to you in my own fashion.

I conceive the first, which may be called the element of fire, as the most subtle and penetrating fluid in the world. And following on from what has been said above concerning the nature of fluid bodies, I imagine its parts to be much smaller and to move much more quickly than any of the parts of other bodies. Or rather, so that I will not have to allow any void in nature, I do not attribute parts having any determinate shape or size to this first element; but I am convinced that the impetuosity of their motion is sufficient to cause it to be divided, in every way and in every sense, by collision with other bodies, and that its parts change shape at every moment to accommodate themselves to the shape of the places they enter. Thus there is never a passage so straight nor an angle so tight among the parts of other bodies that the parts of this element do not enter into it without difficulty and do not fill it entirely.

As for the second, which may be called the element of air, I conceive 25 this too to be a very subtle fluid in comparison with the third, but compared with the first we need to attribute some size and shape to each of its parts and to imagine them as more or less round and joined together like grains of sand or dust. Thus they are not able to arrange themselves or press against each other in such a way that there never remain many small gaps around them; and it is much easier for the first element to slide into these than for the parts of the second to change shape expressly in order to fill them. And so I am convinced that nowhere in the world can this second element be so pure that there is not always a little of the first matter with it.

Beyond these two elements, I accept only a third, namely that of earth. I judge its parts to be proportionately larger than and more slowly

---

[34] There is a representative selection of passages from scholastic texts on the elements in Gilson, *Index*, nos. 156–8.

moving than those of the second, as those of the second are in comparison to those of the first. And indeed I think it is enough to conceive of it
as one or more large masses, whose parts have very little or no motion that
might cause them to change position with respect to one another.

If you find it strange that, in explaining these elements, I do not use the
qualities called 'heat', 'cold', 'moistness', and 'dryness', as the Philosophers
26 do,[35] I shall say that these qualities appear to me to be themselves in
need of explanation. Indeed, unless I am mistaken, not only these four
qualities but all others as well, including even the forms of inanimate
bodies, can be explained without the need to suppose anything in their
matter other than motion, size, shape, and arrangement of its parts.
Because of this, I shall have no difficulty in getting you to understand why
I acknowledge no elements other than the three I have described. For the
difference that must exist between them and those other bodies that the
Philosophers call 'mixed' or 'composite' consists in the fact that the forms
of these mixed bodies always contain in themselves some qualities which
are contrary and counteract one another, or at least do not tend to the
preservation of one another. But the forms of the elements should be
simple and not have any qualities that do not accord with one another so
perfectly that each tends to the preservation of all the others.

Now I cannot find any such forms in the world except the three I have
described. For the form that I have attributed to the first element consists
in its parts moving with such a great speed and being so tiny that there
are no other bodies able to stop them; in addition, they need have no
determinate size, shape, or position. The form of the second element con-
27 sists in its parts having such a middling motion and size that, just as there
are many causes in the world which could increase their motion and
diminish their size, there are as many that could do the opposite; and so
they always remain balanced as it were in the same middling condition.
And the form of the third element consists in its parts being so large or
so closely joined together that they always have the force to resist the
motions of other bodies.[36]

---

[35] On the traditional view of the elements, as represented for example in Aristotle, the four elements
were explained in terms of two pairs of contrary principles: hot versus cold, and wet versus dry.
In this schema, earth was cold and dry, water cold and wet, air hot and wet, and fire hot and dry.

[36] As we shall see when we come to ch. 7, Descartes has what might be described as a 'contest' notion
of collision in which the greater force always 'wins out', rather than a conception in which the
forces are mutually modified. Consequently, a body with the greater force will always be able to
'resist' a lesser one.

Examine as much as you please all the forms that can be given to mixed bodies by the various motions, the various shapes and sizes, and the different arrangement of the parts of matter: I am sure that you will find none that does not contain in itself qualities that tend to bring it about that matter changes and, in changing, to reduce to one of the forms of the elements.

Flame, for example, whose form requires that its parts move very quickly and in addition have some size, as we said above, cannot last long without dying out; for either the size of its parts, in giving them the force to act against other bodies, will cause their motion to diminish, or the violence of their agitation, in causing them to break up on smashing into the bodies they encounter, will cause a diminution of their size. Thus it 28 will be possible for them to be reduced gradually to the form of the third element, or to that of the second, and even some of them to that of the first. In this way, one can see the difference between this flame, or every-day fire, and the element of fire I have described. And you must also recognise that the elements of air and earth – that is, the second and third element – are not more like the gross air we breathe or the earth on which we walk, but that generally all the bodies that appear around us are mixed or composite and subject to decay.

But we do not think therefore that the elements have no places in the world to which they are particularly destined, and where they can be perpetually conserved in their natural purity.[37] On the contrary, each part of matter always tends to one of their forms and, once it has been so reduced, never tends to leave that form. Consequently, even if God had created only mixed bodies at the beginning, all bodies would nonetheless have had the chance to shed their forms and take on those of the elements. Thus we now have every reason to think that all those bodies that are large 29 enough to be counted among the most notable parts of the universe each have the form of one of these elements, and that the only mixed bodies are on the surfaces of these bodies. But there must be mixed bodies, for the elements have quite contrary natures, and two of them could not come

---

[37] In other words, nothing Descartes has argued up to now is contrary to the Aristotelian doctrine of natural place, whereby each of the elements has a natural place to which that element will move if it is unconstrained, and where it will naturally come to rest when it has reached that place. It might seem peculiar that Descartes should revert to such a traditional doctrine here, but the strategy may be to show that construing the elements in terms of size and speed, rather than as being qualitatively different, is still compatible with traditional Aristotelian cosmology. In other words, Descartes could contend that, at this point in the argument, he is merely offering a more economical account of the elements.

into contact without acting against each other's surfaces, and thereby bestowing on the matter there the various forms of these mixed bodies.

In this regard, if we consider in general all the bodies of which the universe is composed, we will find among them only three kinds which can be called large and which can count among the principal parts: namely, the Sun and the fixed stars as the first kind, the heavens as the second, and the Earth with the planets and the comets as the third. That is why we have every reason to think that the Sun and the fixed stars have as their form nothing other than the first element, the heavens the second, and the Earth with the planets and comets the third.

I include the planets and the comets together with the Earth because they, like it, also resist light and reflect its rays, and so I recognise no difference between them. And I include the Sun and the fixed stars together, and attribute to them a nature totally contrary to that of the 30 Earth, because the action of their light is enough for me to recognise that their bodies are of a very subtle and very agitated matter.

As for the heavens, inasmuch as they cannot be perceived by our senses, I think I am right in attributing to them a middle nature between that of the luminous bodies whose action we perceive and that of the solid and heavy bodies whose resistance we perceive.[38]

Finally, we do not perceive mixed bodies anywhere other than on the surface of the Earth.[39] And if we consider that the whole space that contains them – namely that which extends from the highest clouds to the deepest mines that human avarice has ever excavated to extract metals – is extremely small in comparison with the Earth and with the immense expanses of the heavens, we will readily be able to imagine to ourselves that these mixed bodies, taken all together, are just a crust produced on top of the Earth by the agitation and mixing of the matter of the heavens surrounding it.

In this way, we have reason to think that it is not only in the air we breathe, but also in all the other bodies right down to the hardest rocks and the heaviest metals, that there are parts of the element of air mixed with those of earth and consequently parts of the element of fire as well, because they are always found in the pores of the element of air.

[38] As Descartes has already indicated, the heavens are not empty spaces but are filled with a pure air, as distinct from the 'gross' air with which we are familiar on the Earth.

[39] Why Descartes restricts mixed bodies to the Earth is not clear. On the basis of what he has already told us, there is no reason why other planets should not have mixed bodies. It is possible that he associates the presence of mixed bodies with the presence of life, in which case the much-discussed question of 'other worlds' would have been raised, something he may have wanted to avoid.

It should be noted, however, that even though there are parts of these three elements mixed with one another in all bodies, properly speaking 31 only those that can be ascribed to the third element, because of their size or the difficulty they have in moving, compose all the bodies we see around us. For the parts of the other two elements are so subtle that they cannot be perceived by our senses. One may picture all these bodies as sponges in that, even though a sponge has many pores or small holes which are always full of air or water or some similar fluid, we do not think that these fluids enter into its composition.

Many other things remain for me to explain here, and for my own part I would be happy to add a number of other arguments to make my opinions more plausible. But so as to make this long discourse less boring for you, I want to wrap up part of it in the guise of a fable, in the course of which I hope the truth will not fail to manifest itself sufficiently clearly, and that this will be no less pleasing to you than if I were to set it forth wholly naked.

Chapter 6

Description of a new world, and the qualities of the matter of which it is composed[40]

For a while, then, allow your thought to wander beyond this world to view another, wholly new, world, which I call forth in imaginary spaces before it. The Philosophers tell us that these spaces are infinite, and they should certainly be believed, since it is they themselves who invented them.[41] But 32 in order to keep this infinity from impeding and hampering us, let us not try to go all the way, but rather enter it only far enough to lose sight of all the creatures that God made five or six thousand years ago,[42] and after stopping there in some definite place, let us suppose that God creates

---

[40] The heading in the 1664 edition is: *Description of a New World, very easy to know, but nevertheless similar to ours, and even to the chaos which the poets imagined to have preceded it.*

[41] Descartes wrote to Mersenne on 18 December 1629 asking 'whether there is anything definite in religion concerning the extension of created things, that is, whether it is finite or infinite, and whether there are real created bodies in what is called imaginary space, for although I have been afraid to touch on this question, I believe that I shall have to go into it'. We do not have Mersenne's reply, but the question of 'imaginary spaces' was a theologically vexed question because of its connection with the issue of a plurality of worlds. The medieval discussion of the plurality of worlds had focused on a number of different cases, one of which was whether it was possible for there to be a world completely outside this one, that is, outside our cosmos, which existed in an 'imaginary' space.

[42] That is, from the date of the creation of the world as commonly reckoned, on the basis of biblical chronology, in Descartes' time.

anew so much matter all around us that, in whatever direction our imagination may extend, it no longer perceives any place that is empty.

Even though the sea is not infinite, those who are on a vessel in the middle of it can extend their view seemingly to infinity, and nevertheless there is still water beyond what they see. Thus even though our imagination seems to be able to stretch to infinity, and we do not assume this new matter to be infinite, we can assume nevertheless that it fills spaces much greater than those we have imagined. And in order that there be nothing in this assumption that you find objectionable, let us not allow our imagination to extend as far as it could, but purposely confine it to a determinate space which is no greater, say, than the Earth and the principal stars in the firmament, and let us suppose that the matter which God has created extends indefinitely far beyond in all directions. For it is much more reasonable to – and we are much better able to – prescribe limits to the action of our mind than to the works of God.[43]

Now since we are taking the liberty of imagining this matter as we fancy, let us attribute to it, if we may, a nature in which there is absolutely nothing that everyone cannot know as perfectly as possible. To this end, let us explicitly assume that it does not have the form of earth, fire, or air, or any other more specific form, like that of wood, stone, or metal; nor does it have the qualities of being hot or cold, dry or moist, light or heavy, or of having any taste, odour, sound, colour, light, or of any other quality in nature of which there might be said to be something which is not known clearly by everyone.

On the other hand, let us not think that this matter is the 'prime matter' of the Philosophers, which they have stripped so thoroughly of all its forms and qualities that nothing remains in it which can be clearly understood.[44] Let us rather conceive of it as a real, perfectly solid body, which uniformly fills the entire length, breadth, and depth of this great space in the midst of which we have brought our mind to rest. Thus, each

---

[43] In other words, Descartes holds that an infinitely extended universe is within God's power, but he is happy to assume here that his imagined world is simply spatially indefinite. This is a distinction that he will later claim, in conversation with Burman, to have been the first to formulate (AT v. 167).

[44] On the doctrine of 'prime matter', what results when one strips matter of all properties and forms is a propertyless substratum, which Aristotle himself seems to have conceived as a limiting case which could never actually be achieved (in principle), but which later thinkers took to be a genuine substratum underlying forms and qualities. Descartes does not want to conceive of his world in these terms if for no other reason than that he does not want to allow that a world stripped of the forms and qualities he mentions would be propertyless: on the contrary, there is a presumption that in removing these, what we would be left with would be its genuine properties.

of its parts is so proportional to its size that it could not fill a larger one nor squeeze itself into a smaller one; nor, while it remains there, could it allow another body to find a place there.[45]

Let us add further that this matter may be divided into as many parts 34 and shapes as we can imagine, and that each of its parts can take on as many motions as we can conceive. Let us also suppose that God does divide it into many such parts, some larger some smaller, some of one shape some of another, as it pleases us to imagine them. It is not that He separates these parts from one another so that there is some void in between them; rather, let us think of the differences that He creates within this matter as consisting wholly in the diversity of the motions He gives to its parts. From the first instant of their creation, He causes some to start moving in one direction and others in another, some faster and others slower (or even, if you wish, not at all);[46] and He causes them to continue moving thereafter in accordance with the ordinary laws of nature.[47] For God has established these laws in such a marvellous way that even if we suppose that He creates nothing more than what I have said, and even if He does not impose any order or proportion on it but makes it of the most confused and muddled chaos that any of the poets could describe, the laws of nature are sufficient to cause the parts of this chaos to disentangle themselves and arrange themselves in such a good order that they will have the form of a most perfect world, a world in which one 35 will be able to see not only light, but all the other things as well, both general and particular, that appear in the actual world.

But before I explain this at greater length, pause again for a minute to consider this chaos, and note that it contains nothing which you do not know so perfectly that you could not even pretend to be ignorant of it. For the qualities that I have placed in it are only such as you could imagine. And as far as the matter from which I have composed it is concerned, there is nothing simpler or more easily grasped in inanimate creatures.

---

[45] Note that this matter fills a pre-existing space, so cannot be identified with the space, as in the doctrine of corporeal extension that we more usually associate with Descartes. This could be due either to the exigencies of exposition, or to Descartes' not having formulated this doctrine, or at least not having formulated it precisely, at this stage in his thinking. As we shall see, he subsequently goes on to talk of corporeal extension as constitutive of space, and this seems to be a conceptual rather than an empirical point.

[46] There is a hint here that rest is being treated as a limiting case of motion, as opposed to the traditional Aristotelian view whereby rest and motion are treated as qualitatively different states.

[47] The term 'ordinary' laws of nature here indicates that God does not need to act in a special, or exceptional, or miraculous way to bring about His will.

The idea of that matter is such a part of all the ideas that our imagination can form that you must necessarily conceive of it, or you can never imagine anything at all.

Nevertheless, the Philosophers are so subtle that they can find problems in things that seem extremely clear to other men, and the memory of their 'prime matter', which they acknowledge to be rather hard to conceive, may divert them from knowledge of the matter of which I speak. Thus I should say to them at this point that, unless I am mistaken, the whole difficulty they face with their matter derives only from their wanting to distinguish it from its own proper quantity and from its outward extension, that is, from the property it has of occupying space. In this, however, I am willing for them to think they are right, for I have no intention of pausing to contradict them. And they should not find it

36 strange that the quantity of the matter that I have described does not differ from its substance any more than number differs from the things numbered. Nor should they find it strange if I conceive of its extension, or the property it has of occupying space, not as an accident, but as its true form and essence; for they cannot deny that it is quite easy to conceive of it in this way. And my purpose, unlike theirs, is not to explain the things that are in fact in the actual world, but only to make up as I please a world in which there is nothing that the dullest minds cannot conceive, and which nevertheless could not be created exactly the way I have imagined it.

Were I to put in this new world the least thing that is obscure, this obscurity might well conceal some hidden contradiction I had not perceived, and thus without thinking I might suppose something impossible. Instead, since everything I propose here can be imagined distinctly, it is certain that even if there were nothing of this sort in the old world, God can nevertheless create it in a new one; for it is certain that He can create everything we imagine.

Chapter 7

The Laws of Nature of this new world[48]

But I do not want to delay any longer telling you the means by which Nature alone is able to untangle the confusion of the chaos which I have

---

[48] The heading in the 1664 edition is: *By what Laws and by what Means the parts of this World will extricate themselves, by themselves, from the Chaos and Confusion they were in.*

been speaking about, and what the Laws of Nature that God has imposed on it are.

Take it then, first, that by 'Nature' here I do not mean some deity or 37 other sort of imaginary power. Rather, I use the word to signify matter itself, in so far as I am considering it taken together with the totality of qualities I have attributed to it, and on the condition that God continues to preserve it in the same way that He created it. For it necessarily follows from the mere fact that He continues to preserve it thus that there may be many changes in its parts that cannot, it seems to me, properly be attributed to the action of God, because this action never changes, and which I therefore attribute to Nature. The rules by which these changes take place I call the Laws of Nature.

In order to understand this better, remember that among the various qualities of matter we have supposed that its parts have had various different motions since the moment they were created, and furthermore that they all touch one another on all sides, without there being any void in between any two of them. From this it follows necessarily that from the time they begin to move, they also begin to change and diversify their motions by colliding with one another. Thus, while God subsequently preserves them in the same way He created them, He does not preserve them in the same state. That is to say, if God always acts in the same way and consequently always produces substantially the same effect, many 38 differences in this effect occur, as if by accident. And it is easy to accept that God, who is, as everyone must know, immutable, always acts in the same way. Without my going any further into these metaphysical considerations, however, I will set out here two or three of the principal rules by which we must believe God to cause the nature of this new world to act, and these will be enough, I believe, to acquaint you with all the others.[49]

The first is that each particular part of matter always continues in the same state unless collision with others forces it to change its state. That is to say, if the part has some size, it will never become smaller unless others divide it; if it is round or square, it will never change that shape unless others force it to; if it is brought to rest in some place it will never depart from that place unless others drive it out; and if it has once begun

---

[49] The expression 'all the others' here presumably refers to the laws of collision, which Descartes does not set out here but presents for the first time in print fifteen years later, in his *Principles of Philosophy*.

to move, it will always continue with an equal force until others stop or retard it.[50]

There is no one who does not believe that this same rule is observed in the old world[51] as regards size, shape, rest, and a thousand other things. But the Philosophers have exempted motion from it, which is the one thing that I most explicitly wish to include. Do not think that I intend to contradict them, though: the motion that they speak of is so very different from that which I conceive that it can easily happen that what is true of the one is not true of the other.

They themselves admit that the nature of their motion is very little understood. And trying to make it more intelligible, they have still not been able to explain it more clearly than in these terms: *Motus est actus entis in potentia, prout in potentia est.*[52] These terms are so obscure to me that I am compelled to leave them in Latin because I cannot interpret them. (And in fact the words 'motion is the act of a being which is in potency, in so far as it is in potency' are no clearer for being in the vernacular.) By contrast, the nature of the motion that I mean to speak of here is so easily known that even geometers, who among all men are the most concerned to conceive the things they study very distinctly, have judged it simpler and more intelligible than the nature of surfaces and lines, as is shown by the fact that they explain 'line' as the motion of a point and 'surface' as the motion of a line.[53]

The Philosophers also posit many motions which they believe can occur without any body's changing place, such as those they call *motus ad formam, motus ad calorem, motus ad quantitatem* (motion with respect to form, motion with respect to heat, motion with respect to quantity) and

---

[50] Note that no direction of motion is specified here: as it stands, the formulation is compatible with rectilinear motion, circular motion, parabolic motion, and so on.

[51] That is, the world of common sense, as it would have been described by most of Descartes' contemporaries.

[52] Aristotle, *Physics* 201ª10. Compare Descartes' criticism of this definition in the *Rules* (AT x. 426; CSM i. 49).

[53] These accounts/definitions in terms of motion are not to be found in the three best-known ancient geometers – Euclid, Archimedes, and Apollonius. Euclid, for example, in the definitions at the beginning of the *Elements*, simply defines a line as a breadthless length and a surface as something having only length and breadth (cf. Aristotle, *Metaphysics* 1016ᵇ24–31). The geometers of antiquity would not have denied that a line can be generated from the motion of a point, and various constructions, such as Pappus' method for generating a quadratrix, rely on this property of moving points. Nevertheless, unless I have missed some crucial passage, it is not true that the classical geometers *explain* a line as the motion of a point. It is possible that Descartes is thinking of something in a contemporary geometrical text, but it is also possible that he is providing his own gloss on geometrical practice.

countless others.[54] For my own part, I know of no motion other than 40
that which is easier to conceive of than the lines of geometers, by which
bodies pass from one place to another and successively occupy all the
spaces in between.

In addition, the Philosophers attribute to the least of these motions a
being much more solid and real than they do to rest, which they say is
merely a privation of motion. For my part, I conceive of rest as a quality
also, which should be attributed to matter while it remains in one place,
just as motion is a quality attributed to matter while it is changing place.

Finally, the motion of which they speak has a very strange nature in
that all other things have as a goal their perfection, and strive only to
preserve themselves, whereas it has no other end or goal than rest, and
contrary to all laws of nature it strives of itself to destroy itself. By
contrast, the motion I suppose follows the same laws of nature as do
generally all the dispositions and qualities found in matter. This includes
those that the Schoolmen call *modos et entia rationis cum fundamento in re*
(modes and beings of thought based in the thing) as well as those they call
qualitates reales (their real qualities), in which I frankly confess I cannot
find any more reality than in the others.

I put forward as my second rule that when one of these bodies pushes 41
another it cannot give the other any motion except by losing as much of
its own motion at the same time; nor can it take away any of the other's
motion unless its own is increased by the same amount. This rule,
together with the preceding, accords very well with all those observations
in which we see one body begin or cease to move because it is pushed
or stopped by another. For, having assumed the previous rule, we are
free from the difficulty in which the Schoolmen find themselves when
they wish to explain why a stone continues to move for some time after
leaving the hand of the person who threw it. For we should ask instead,
why does the stone not continue to move forever? Yet the reason is easy
to give. For who can deny that the air in which it is moving offers it some
resistance? We hear it whistle when it divides the air, and if a fan, or some
other very light and extended body, is moved through the air, we shall
even be able to feel by the weight in our hand that the air is impeding its

---

[54] The Latin term *motus* translates the Greek term *kinesis*, which refers to the general category of
change (excluding generation and corruption). Descartes' somewhat disingenuous presentation
of the point obscures the fact that his claim is a contentious one: namely that there is no form of
change other than local motion.

motion rather than keeping it moving, as some have wanted to say.[55] Now suppose we refuse to explain the effects of the air's resistance in line with our second rule, thinking that the more a body can resist the more it is capable of stopping the motion of others, as we might initially be per-

42 suaded perhaps. We will then have great difficulty explaining why the motion of this stone is diminished more in colliding with a soft body which offers moderate resistance than when it collides with a harder body which resists it more. Likewise, we shall find it hard to explain why, as soon as it has exerted itself a little against the latter, it immediately turns around, rather than stopping or interrupting its motion. But if we accept this rule, there is no difficulty here at all. For it tells us that the motion of one body is not retarded by its collision with another in proportion to how much the latter resists it, but only in proportion to how much the latter's resistance is surmounted, and to the extent that, in obeying the law, it receives into itself the force of motion that the former gives up.

Now although, in most of the motions we see in the actual world, we cannot perceive that the bodies that begin or cease to move are pushed or stopped by some others, we have no reason to judge that these two rules are not being followed exactly. For it is certain that such bodies can often receive their agitation from the two elements of air and fire, which are always found among them without being perceptible (as has just been said), or that they may receive it from the ordinary air, which also cannot be perceived by the senses. It is certain too that they can transfer this agitation sometimes to the grosser air, and sometimes to the whole mass of the earth; and when dispersed therein, it also cannot be perceived.

43 But even if everything our senses ever experienced in the actual world seemed manifestly contrary to what is contained in these two rules, the reasoning that has taught me them seems so strong that I cannot help believing myself obliged to suppose them in the new world that I am describing to you, for what more firm and solid a foundation could one find to establish a truth, even if one wished to choose it at will, than the very firmness and immutability which is in God?

Now these two rules follow manifestly from the sole fact that God is

---

[55] What Descartes has in mind here is the standard Aristotelian account whereby the continued ('violent') motion of a projectile is effected by the surrounding air, which pushes the projectile onwards. Although this account was defended in the Coimbra commentaries (see Gilson, *Index*, item 300), many late Scholastic thinkers favoured the *impetus* theory, whereby the continued motion is explained in terms of the transfer of a force or motion from the hand to the projectile when it is thrown, a force or motion that gradually dies down as the projectile rises.

immutable and that, acting always in the same way, He always produces the same effect. For on the assumption that He placed a certain amount of motion in matter in general at the first instant He created it, we must admit either that He preserves the same amount of motion in it, or not believe that He always acts in the same way.[56] If we assume, in addition, that from this first instant the various parts of matter, in which these motions are found unequally dispersed, began to retain them or transfer them from one to another, according as they had the force to do so, then we must of necessity hold that God causes them to continue always doing so. And that is what these two rules specify.

I shall add as a third rule that, when a body is moving, even if its motion most often takes place along a curved line and, as we said above, it can 44 never make any movement that is not in some way circular, nevertheless each of its parts individually tends always to continue moving along a straight line. And so the action of these parts, that is the inclination they have to move, is different from their motion.

For example, if we make a wheel turn on its axle, even though its parts go in a circle (because, being joined to one another, they cannot do otherwise), nevertheless their inclination is to go straight ahead, as appears clearly if one of them is accidentally detached from the others, for as soon as it is free its motion ceases to be circular and continues in a straight line.

By the same token, when a stone is swung in a sling, not only does it fly straight out when it leaves the sling, but while it is in the sling it presses against the middle of it[57] and causes the cord to stretch. This shows clearly that it always has a tendency to go in a straight line and that it goes in a circle only under constraint.[58]

This rule rests on the same foundation as the other two, and depends solely on God's conserving everything by a continuous action, and

---

[56] The assumption that acting in the same way always produces the same effect is clearly crucial here, for if it were abandoned one could allow both that God always acts in the same way and yet does not preserve the same amount of motion. It's far from clear that Aristotle, for one, would have accepted this principle.

[57] The term 'middle' here might be a little misleading at first. We have to assume that the sling is a length of material – joined by a cord at both ends – which is doubled over so that the stone lies in the fold. The stone presses outward at this fold, and not of course toward the middle of the circle through which it is swung.

[58] This seems as unambiguous a statement of the theory that only rectilinear motion is inertial as one could wish for. But matters are not so straightforward. As we shall see below, in ch. 13 Descartes argues in a way that suggests that not just rectilinear motion, but, at least in some circumstances, circular motion can also be treated as inertial.

consequently on His conserving it not as it may have been some time earlier but precisely as it is at the very instant He conserves it. So, of all
45 motions, only motion in a straight line is entirely simple and has a nature which may be grasped wholly in an instant. For in order to conceive of such motion it is enough to think that a body is in the process of moving in a certain direction, and that this is the case at each determinable instant during the time that it is moving. By contrast, to conceive of circular motion, or any other possible motion, it is necessary to consider at least two of its instants, or rather two of its parts, and the relation between them.[59] But so that the Philosophers (or rather the Sophists) do not find the opportunity here to engage in their useless subtleties, note that I am not saying that rectilinear motion can take place in an instant; but only that all that is required to produce it is found in bodies in each instant that may be determined while they are moving, whereas not everything that is required to produce circular motion is present.

For example, suppose a stone is moving in a sling along the circle marked AB [fig. 1], and consider it exactly as it is at the instant it arrives at the point A. You will readily find that it is in the process of moving, for it does not stop there, and that it is moving in a certain direction, namely
46 towards C, for it is in that direction that its action is directed in that instant. But nothing can be found here that makes its motion circular. Thus, supposing that the stone then begins to leave the sling and that God continues to preserve it as it is at that moment, it is certain that He will not preserve it with the inclination to travel in a circle along the line AB, but with the inclination to travel straight ahead toward point C.

According to this rule, then, we must say that God alone is the author of all the motions in the world in so far as they exist and in so far as they are straight, but that it is the various dispositions of matter that render the motions irregular and curved. Likewise, the theologians teach us that
47 God is also the author of all our actions, in so far as they exist and in so far as they have some goodness, but that it is the various dispositions of our wills that can render them evil.

[59] The argument here relies on motion being conceived in terms of a discontinuous series of instants, for if it is conceived of as something continuous then the requisite distinction between rectilinear and circular motion cannot be made. In Aristotelian accounts, motion, and change more generally, would have been conceived of in terms of a continuous process: and motion thought of teleologically would always be so conceived. Descartes' commitment to the discontinuous nature of motion derives above all from his early work in hydrostatics, where, because in statics one is concerned to describe states of equilibrium, it is instantaneous tendencies to motion, rather than motion proper, that is the central concern.

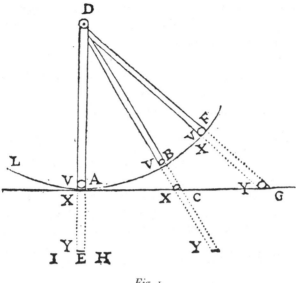

Fig. 1

I could set out many further rules here for determining in detail when and how, and by how much, the motion of each body can be diverted and increased or decreased by colliding with others, that is, rules that comprise all the effects of nature in a summary way.[60] But I shall be content to tell you that, apart from the three laws that I have explained, I wish to suppose no others but those that most certainly follow from the eternal truths on which mathematicians have generally supported their most certain and most evident demonstrations: the truths, I say, according to which God Himself has taught us He disposed all things in number, weight, and measure.[61] The knowledge of these truths is so natural to our souls that we cannot but judge them infallible when we conceive them distinctly, nor doubt that if God had created many worlds, they would be as true in each of them as in this one. Thus those who know how to examine the consequences of these truths and of our rules sufficiently

---

[60] The reference here is to the rules of collision of the kind set out later in the *Principles of Philosophy* Part II, arts. 45–62. We do not know whether Descartes had formulated these rules by the time of writing the present treatise.

[61] In the notes to his translation of *Le Monde*, Mahoney notes that this statement was a commonplace of medieval thought, and formed the opening line of Sacrobosco's *Algorismus vulgaris*, the standard arithmetic textbook from the mid-thirteenth to the mid-sixteenth centuries. (M. S. Mahoney, *René Descartes, Le Monde ou Traité de la Lumière* (New York, 1979), 220 n. 29.)

will be able to recognise effects by their causes. To express myself in scholastic terms, they will be able to have *a priori* demonstrations of everything that can be produced in this new world.[62]

48 And so that there will be nothing to prevent this, we shall, if you please, assume in addition that God will never perform a miracle in the new world, and that the intelligences, or rational souls, which we might later suppose to be there, will not disrupt the ordinary course of nature in any way.

Nevertheless, after this, I do not promise to set out exact demonstrations of everything I say. It will be enough for me to open up the way for you to find them yourselves, when you take the trouble to look for them. Most minds lose interest when one makes things too easy for them. And so as to present a picture which pleases you here, I must use shading as well as bright colours. So I shall be content to continue with the description I have begun, as if my intention were simply to tell you a fable.

Chapter 8

On the formation of the sun and the stars in this new world[63]

Whatever inequality and confusion we might suppose God to have put among the parts of matter at the beginning, following the laws He imposed on Nature, the parts of matter must subsequently almost all have been reduced to one size and to one moderate motion and thus have taken the form of the second element, following the explanation that I gave

49 above. For when we consider this matter in the state that it could have been in before God started to move it, we must imagine it as the hardest and most solid body in the world. And, since one would not be able to push any part of such a body without pushing or pulling all the other parts in the same operation, so we must imagine that the action or force of moving or dividing, which, being placed first in some parts of matter, spread out and distributed itself in all the others at the same instant, as equally as it could.

It is true that this equality could not be completely perfect. For, first,

---

[62] 'A priori' here means simply from more basic principles, and has no connotations about whether the demonstration involved would be empirical or not: this modern connotation of the term 'a priori' is due to Leibniz and above all to Kant.

[63] The heading in the 1664 edition is: *How, in the world as described, the heavens, the sun and the stars are formed.*

because there is no void at all in this new world, it was not possible for all the parts of matter to move in a straight line. Rather, since they were all just about equal and as easily divisible, they all had to form together into various circular motions. And yet, because we suppose that God initially moved them in different ways, we should not imagine that they all came together to turn around a single centre, but around many different ones, which we may imagine to be variously situated with respect to one another.

Consequently, we can conclude that they must have been naturally less agitated and smaller, or both, at those places nearest to these centres than at those farthest way. For since all of them have an inclination to continue their motion in a straight line, it is certain that the strongest – that is, the 50 largest among those that are equally agitated, and the most agitated among those that are equally large – had to describe the largest circles, that is, those circles that approach a straight line most closely. And as for the matter contained between three or more of these circles, it could initially have been much less divided and less agitated than all the rest. And what is more, especially since we suppose that God initially placed every kind of inequality among the parts of this matter, we must imagine that there were then all sorts of sizes and shapes, and dispositions to move and not to move, in all ways and in all directions.

But this does not prevent them from having subsequently all been made fairly equal, especially those that remained an equal distance from the centres around which they were turning. For since some could not move without the others moving, the more agitated had to communicate some of their motion to those that were less so, and the larger had to break up and divide so as to be able to pass through the same places as those that went before them, or so that they might rise higher. And thus all the parts were soon arranged in order, each being more or less distant from the centre around which it had taken its course, according as it was more or less large and agitated compared to the others. Indeed, inasmuch as size 51 always resists speed of motion,[64] one must imagine that the parts more distant from each centre were those which, being somewhat smaller than the ones closer to the centre, were thereby very much more agitated.

Exactly the same holds for their shapes, for even if we were to suppose that there were initially all kinds of shapes, and that they had for the most

[64] It is unclear here what 'size' ('grosseur') is; it could be volume, surface area, weight, or whatever.

part many angles and many sides, like the pieces that fly off from a stone when it is broken, it is certain that subsequently, in moving and hurtling themselves against one another, they gradually had to break the small points of their angles and dull the square edges of their sides, until they had almost all been rounded off, just as grains of sand and pebbles do when they roll with the water of a river. Thus there cannot now be any considerable difference between those parts that are reasonably close together, nor indeed even among those that are quite distant, except that the one can move a bit more quickly than the other and be slightly larger or smaller. And this does not prevent our attributing the same form to all of them.

But an exception must be made for some which, having initially been very much larger than the others, could not be so easily divided, or which, 52 having had shapes which were very irregular and prevented [this], joined together severally rather than breaking up and rounding off. These have consequently retained the form of the third element and served to make up the planets and the comets, as I shall tell you later.

We must also note in connection with the matter that emerged from around the parts of the second element that, to the extent that these broke and dulled the small points on their angles in the course of rounding off, it necessarily had to acquire a very much faster motion than they, and along with this a facility for dividing and changing shape at every moment so as to accommodate itself to the shape of the places where it found itself. And so it took the form of the first element.

I say that it had to acquire a much faster motion than theirs. The reason for this is clear. For, having to go off to the side, and through very narrow passages, out of the small spaces left between the parts as they proceeded to collide head-on with one another, it had to traverse a very much greater path than they had in the same time.

We must also note that what remains of the first element – over and above what is needed to fill the small spaces that the parts of the second [element], which are round, necessarily leave around them – must move back towards the centres around which those parts turn, because [the 53 latter] occupy all the other, more distant places. At [these centres], the remaining element must compose perfectly fluid and subtle round bodies which, because they incessantly turn very much more quickly than and in the same direction as the parts of the second element surrounding them, have the force to increase the agitation of those parts to which they

are closest and even – in moving from the centre towards the circumference – to push the parts in every direction, just as they push one another;
and this occurs through an action that I must soon explain as precisely as
I can. For I warn you here in advance that it is this action that we shall
take to be light, just as we shall take one of those round bodies composed
of nothing but the matter of the first element to be the Sun, and the
others to be the fixed stars, of the new world I am describing to you; and
we shall take the matter of the second element turning around them to be
the heavens.

Imagine, for example, that the points s, e, ε, and a are the centres of
which I speak [fig. 2], that all the matter contained in the space FGGF is a
heaven turning about the Sun marked s, that all the matter of the space
HGGH is another heaven turning about the star marked ε, and so on for the
others. So that there are as many different heavens as there are stars, and
since the number of stars is indefinite so too is the number of heavens.
And the firmament is just a surface without thickness, separating all the 54
heavens from one another.

Imagine also that the parts of the second element in the neighbourhood
of F, or G, are more agitated than those in the neighbourhood of K, or L,
so that their speed gradually decreases from the outside circumference of
each heaven to a particular place – such as to the sphere KK about the Sun,
for example, and to the sphere LL about the star ε – and then increases
gradually from there to the centres of the heavens because of the agitation of the stars located there. Thus, while the parts of the second element
in the neighbourhood of K have the opportunity to describe a complete
circle there about the Sun, those in the neighbourhood of T, which I am
supposing to be ten times closer, not only have the opportunity to
describe ten circles, which they would do if they moved only at the same
speed, but perhaps more than thirty. And again, those parts towards F, or
towards G, which I am supposing to be two or three thousand times more
distant, can perhaps describe more than sixty circles. From this you will
realise immediately that the highest planets must move more slowly than
the lowest, that is, those closest to the Sun, and that all the planets move
together more slowly than the comets, which are nevertheless further
away.

As for the size of each of the parts of the second element, we can 55
imagine it to be equal among all those between the outside circumference
FGGF of the heaven and the circle KK, or even that the highest amongst 56

35

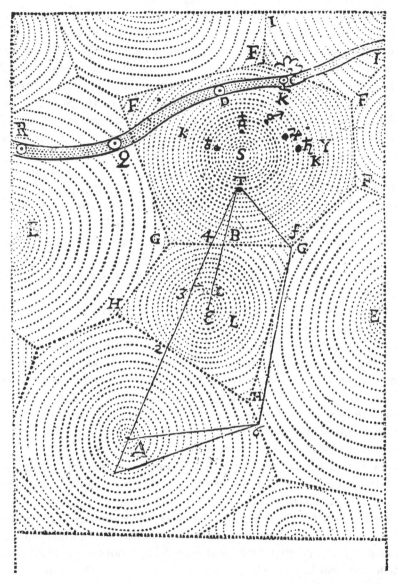

Fig. 2

them are slightly smaller than the lowest, provided that it is not supposed that the difference in their sizes is proportionately greater than that of their speeds. From the circle κ to the Sun, by contrast, the lowest parts must be taken to be the smallest, and the difference in size must be taken to be proportionately greater than, or at least in proportion to, that of their speeds. For otherwise, since those lowest parts, owing to their agitation, are the strongest, they would move to occupy the place of the highest.

Note finally that, given the way in which I have said that the Sun and the other fixed stars were formed, their bodies can be so small in relation to the heavens containing them that even all the circles KK, LL, etc., which mark the extent to which the agitation of these bodies advances the course of the matter of the second element, can be considered merely as the points that mark their centre. In the same way, the new astronomers all but consider the whole sphere of Saturn to be a point in comparison with the firmament.

Chapter 9

On the origin and the course of the planets and comets in general, and of comets in particular[65]

In order for me to begin to tell you about the planets and comets, consider that, given the diversity in the parts of matter that I have supposed, even though most of them have – through breaking up and dividing as a result of collision with one another – taken the form of the first and second element, there nevertheless remains to be found among them two kinds 57 that had to retain the form of the third element. These are those whose shapes were so extended and were sufficiently able to prevent this to such a degree that, when they collided with one another, it was easier for several of them to join together, and in this way to become larger rather than breaking up and becoming smaller; and those which, having been the largest and most massive of all from the very start, were well able to break and shatter the others by striking them, but which were not in turn broken or shattered themselves.[66]

Now irrespective of whether you imagine these parts to have been

---

[65] The heading in the 1664 edition is: *On the origin, the course and other properties of the comets and the planets in general, and of comets in particular.*

[66] Here we see one of the more important consequences of Descartes' 'contestant' notion of forces, whereby the larger or stronger body always 'wins out' in a collision, changing the state of the body it collides with but itself remaining unaffected. Unless this happened, Descartes would not be able to explain the formation or continuance of third element bodies.

initially agitated very much or very little, it is certain that they subsequently had to move with the same agitation as the matter of the heaven that contained them. For if their initial motion was quicker than that of this matter, since they would not have been able to avoid pushing it when it was in their path and they collided with it, in a short time they would have to transfer part of their agitation to it. On the other hand, if they had in themselves no inclination to move, because they were surrounded on all sides by this celestial matter they would nonetheless necessarily have to follow its course, just as we constantly see that boats and other kinds of body that float on water – the biggest and most bulky as well as those

58 that are less so – follow the course of the water they are in when there is nothing to prevent them from doing so.

And note that, among the many different bodies that float thus on the water, those that are rather big and bulky – as boats usually are, especially the largest and most heavily laden – always have much more force than the water to continue their motion, even though it is solely from the water that they received their motion. By contrast, floating bodies that are very light, such as the lumps of white scum that one sees floating along the shores during storms, have less force to continue moving. If you imagine, then, two rivers that join together at some point and separate shortly afterwards before their waters – which we must assume to be very calm and to be of roughly equal force, but also to be very rapid – have had an opportunity to mix, the boats and other massive and heavy bodies that are borne by the course of the one river will be able to pass easily into the other river, whereas the lightest bodies will swerve away from it and will be thrown back by the force of the water towards wherever it is the least rapid.

For example, if ABF and CDG are two rivers [fig. 3] which, coming from different directions, meet at E and then turn away, AB going toward F and CD toward G, then it is certain that boat H, following the course of the river AB, must pass through E toward G, and reciprocally boat I towards F, unless

59 both meet at the point of intersection at the same time, in which case the larger and stronger will break the other. By contrast, scum, leaves of trees, feathers, straw, and other such light bodies which might be floating at A must be pushed by the course of the water containing them, not towards E and G, but toward B, where we must consider the water to be less strong and rapid than at E, since there [at B] it takes a course along a line which is not as close to a straight one.

38

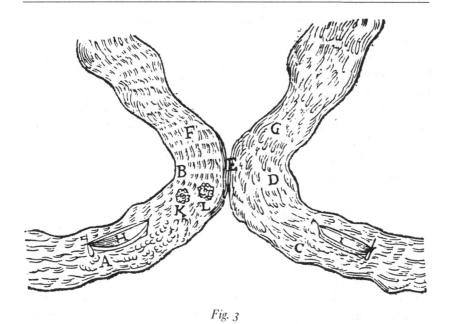

Fig. 3

Moreover, we must consider that not just these light bodies, but also heavier and bulkier ones, can be joined together upon meeting and that, turning then with the water that bears them along, several 60 together can compose large balls such as you see at K and L; some of these, such as L, go toward E and others, such as K, toward B, depending on their degree of solidity and on the size and massiveness of their parts.

From this example, it is easy to understand that, no matter where the parts of matter that could not take the form of the second or the first element may have been initially, the larger and more bulky of them soon had to take their course toward the outer circumference of the heaven that contained them, and subsequently pass from one into another of these heavens without ever remaining long in the same heaven. The less massive ones, on the other hand, had each to be pushed toward the centre of the heaven containing it by the course of the matter in that heaven. And given the shapes that I attributed to them, when they collided with one another they had to join together and compose large balls which, turning in the heavens, have a motion which is tempered by all the motions their individual parts would have, had they been separate.

39

So, some tend to move toward the circumference of these heavens, and others toward their centres.

I would also have you bear in mind that we should take those that range toward the centre of any heaven to be the planets and those that pass 61 across different heavens to be comets.

Now first, as regards these comets, it should be noted that there has to be few of them in this new world, compared to the number of heavens. For even if there were initially many of them, in passing across the heavens over the course of time almost all of them would have collided with one another and broken one another up, just as I have said the two boats do when they meet, so that now only the largest would remain.

It must also be noted that, when they pass in this way from one heaven to another, they always push in front of them a bit of matter from the heaven they are leaving, and they remain enveloped by it for some time until they have entered far enough within the limits of the other heaven. Once there, they shed it almost all at once, taking less time, perhaps, than the Sun does to rise on our horizon in the morning. Because of this, they move much more slowly when they tend to leave a heaven than they do soon after having entered it.

For example, you can see [fig. 2] that the comet that travels along the 62 line CDQR, having already entered quite deep within the limits of the heaven FG, still remains at point C enveloped by matter from heaven FI, from which it comes, and cannot be entirely rid of that matter before it is in the vicinity of point D. But as soon as it has arrived there it begins to follow the course of heaven FG and thus to move much faster than it did before. Then, continuing in its course from there toward R, its motion must again gradually slow down proportionately to its approach to point Q, both because of the resistance of the heaven, whose boundaries it is beginning to enter within, and because, since there is less distance between s and D than between s and Q, all the matter of the heaven between s and D, where the distance is smaller, moves more quickly there; just as we observe rivers always flowing more swiftly in those places where their bed is narrower and more confined than in those where it is wider and more extensive.

Moreover, we must note that this comet should be visible to those who live at the centre of the heaven FG only during the time it takes to pass from D to Q, something that will soon be clearer to you, when I have told you what light is. And in the same way, you will grasp that its motion should appear to observers to be much faster, its size much greater, and its light far clearer, at the first moment they see it than towards the end. 63

And as well as this, if you consider carefully the way that the light from the comet must be spread out and distributed in all directions in the heaven, you will also be readily able to understand that, being very large, as we must suppose it to be, it is possible that there may appear around it certain rays that sometimes extend in the form of a halo on all sides, and sometimes gather in the form of a tail on one side only, depending on where it is observed from.[67] Thus this comet lacks none at all of the properties observed up to now in those that have been seen in the actual world: or at least none of the properties that should be taken as true. For if some historians, in order to provide a miracle warning of the Turkish crescent, tell us that in the year 1450 the Moon was eclipsed by a comet that passed below it, or some such thing; and if astronomers, calculating poorly the amount of refraction of the heavens, which they did not know, and the speed of the comets, which is uncertain, attribute to them enough parallax to be placed among the heavens, or even below them (where some wish to pull them, as if by force), then we are not obliged to believe them.

Chapter 10

Of the planets in general, and in particular of the Earth and the Moon[68]

There are, in the same way, several things to note concerning the planets. The first is that, even though they all tend toward the centres of the 64 heavens containing them, this is not thereby to say that they could ever reach those centres. For as I have already said above, these are occupied by the Sun and the other fixed stars. But so that I can show you distinctly in what places the planets should stop, look for example at the one marked ♄[69] [fig 2], which I suppose to follow the course of the matter of the heaven toward the circle κ, and reflect that, if this planet had the slightest bit more force to continue its motion in a straight line than the parts of the second element surrounding it do, then instead of always following this

---

[67] The question of the rings of light in the formation of parhelia was the original motivation for this treatise, and it is not surprising that Descartes should draw so much attention to the optical aspects of comets.

[68] The heading in the 1664 edition is: *The explanation of the planets and principally of the earth and the moon.*

[69] Saturn.

41

circle K, it would go towards Y, and thus it would be further away than it is from centre S. Then, to the extent that the parts of the second element that surround it at Y move faster and are even a bit smaller than – or at least are no larger than – those at K, they would give it even more force to pass beyond this toward F; in this way, it would go out to the circumference of the heaven, without being able to stop anywhere in between. From there it would then be able to pass into another heaven and thus, rather than being a planet, it would be a comet.

65 Whence you see that no star can stop anywhere in all that vast space between the circle K and the circumference of the heaven FGGF, through which the comets take their course. In addition it is impossible for the planets to have more force to continue their motion in a straight line than the parts of the second element at K, when those planets move with the same agitation along with them; and all those bodies that have more are comets.

Let us imagine, then, that this planet ♄ has less force than the parts of the second element surrounding it, so that those parts that follow it and are positioned slightly lower than it can divert it, and that instead of following circle K, it descends toward the planet marked ♃,[70] and since the planet ♄ is there, it can come about that it is exactly as strong as the parts of the second element that will then surround it. This occurs because, these parts of the second element being more agitated than those at K, they will also agitate the planet more, and being in addition much smaller, they will not be able to offer as much resistance. In the event of this, the planet will remain perfectly balanced in the middle of them and will there take its course in the same direction around the Sun as they do, without there being any variation in distance from the Sun from one time to another, except in so far as they can also vary in distance from it.[71]

But if this planet ♄, being at ♃, still has less force to continue its 66 motion in a straight line than the celestial matter found there, it will be pushed lower still by the matter, towards the planet marked ♂, and so on, until finally it is surrounded by a matter that has neither more nor less force than it.

---

[70] Jupiter.
[71] In telling us here that the shape of the orbit must be the same for all the planets, Descartes is possibly hinting at elliptical orbits. He may have known of Kepler's account of elliptical orbits from a visit to Beeckman in 1628 – Beeckman was studying Kepler at the time – but like Beeckman, he seems quite indifferent to the exact shape of the orbit, apparently assuming that a circle will stand in for any closed curve.

Thus you see that there can be different planets, at varying distances from the Sun, such as ♄, ♃, ♂, T, ♀, ☿.[72] Of these, the lowest and the least massive can reach the Sun's surface, but the highest never pass beyond circle K, which, although it is very large in comparison with each planet taken individually, is nonetheless so small in comparison to the whole heaven FGGF that, as I have already said earlier, it can be considered as its centre.

But if this is still not sufficient demonstration of how it can happen that the parts of the heaven beyond circle K, being incomparably smaller than the planets, still have more force than they to continue in their motion in a straight line, consider that this force does not depend solely on the amount of matter that is in each body, but also on the extent of its surface. For even though, when two bodies move equally fast, it is correct to say that if one contains twice as much matter as the other it also has twice the amount of agitation, this is not to say thereby that it has twice the force to continue to move in a straight line; but rather that it will have exactly twice as much if, in addition, its surface is exactly twice the extent, because it will always meet twice as many other bodies resisting it, and it will have much less force to continue if its surface is much more than twice in extent.[73]

Now you know that the parts of the heaven are more or less completely round and thus that, of all shapes, they have the one that includes the most matter within the least surface, whereas the planets, being composed of small parts which have very irregular and extended shapes, have large surfaces in proportion to the amount of their matter. In this way, it is possible for the planets to have a greater [surface area to volume ratio] than most of these parts of the heaven and yet have a smaller one than some of the smallest parts that are closest to the centres. For it must be

---

[72] These are respectively Saturn, Jupiter, Mars, Earth, Venus, and Mercury. T stands for 'Terre' (Earth); the other symbols are traditional ones, and the Sun, Moon, and the five planets mentioned here were associated with the seven metals, the symbols for the metals and the planets being interchangeable. Note that Descartes seems to be assuming here that Saturn is larger than Jupiter. Tycho Brahe, who provided the most accurate observations before the introduction of the telescope, had estimated that Saturn's volume is 22 times that of the Earth with a radius 31/11 times that of Earth, whereas the volume of Jupiter is only 14 times greater and its radius only 12/5 times that of Earth. Once telescopic observation had established disc sizes, however, it became clear that Jupiter was larger than Saturn.

[73] In other words, a body's 'force' is proportional to its surface area, not to its volume. This does not help us identify what kind of force Descartes has in mind: the analogy with the boats in the river suggests something like momentum, but the appeal to surface area as a measure of the force makes this an unlikely candidate here.

43

understood that, with two completely massive balls, such as the parts of the heavens, the smaller always has more surface in proportion to its quantity than the larger.[74]

All this can be confirmed easily by observation. For if one pushes a large ball composed of several branches of trees haphazardly joined together and piled on top of one another – as we must imagine the parts of matter making up the planets to be – it is certain that, even if pushed
68 by a force entirely proportional to its size, it will not be able to continue as far in its motion as would another ball which was very much smaller and composed of the same wood, but wholly massive. On the other hand, we could of course make another ball of the same wood, and wholly massive, but so extremely small that it would have much less force to continue in its motion than the first had. Finally, it is certain that the first ball can have more or less force to continue its motion depending on the extent to which the branches composing it are large and compressed.

Whence you see how various planets can be suspended within circle K at various distances from the Sun, and how it is not just those that outwardly appear the largest, but those that are the most solid and the most massive inside that must be the most distant.

After this, we must note that, just as we observe that boats following the course of a river never move as fast as the water that bears them, nor indeed do the larger among them move as fast as the smaller, so too, even though the planets follow the course of the celestial matter without resistance, and move with the same agitation as it, that is not to say thereby
69 that they ever move exactly as quickly as it. And indeed the inequality of their motion must bear some relation to the inequality between the size of their mass and the smallness of the parts of the heaven that surround them. This is because, generally speaking, the larger a body the easier it is for it to communicate some of its motion to other bodies, and the more difficult it is for others to communicate to it something of their motion. For although several small bodies all acting together upon a larger one may have as much force as it, nonetheless they cannot make it move as fast in every direction as they do because, if they agree in some of the motions that they communicate to it, they almost certainly differ at the same time in others which they cannot communicate to it.

[74] This follows simply from the fact that the surface area of a sphere varies as the square of its radius, whereas its volume varies as the cube of its radius, and that consequently, as the sphere expands, its surface area increases at a lesser rate than its volume.

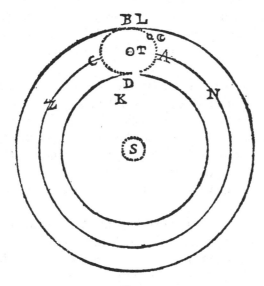

Fig. 4

Now two things follow from this which seem to me to be very significant. The first is that the matter of the heaven must make the planets turn
not only around the Sun, but also around their own centre, except where
there is something particular preventing them from doing so, and consequently that the matter must form around the planets small heavens that
move in the same direction as the greater heaven. The second is that, if
two planets meet that are unequal in size but disposed to take their course
in the heavens at the same distance from the Sun, and if one of them is
exactly as massive as the other is larger, then the smaller of the two, 70
moving more quickly than the larger one, must become joined to the
little heaven around that larger heaven and turn continually around it.

For since the parts of the heaven that are, say, at A [see fig. 4] move faster
than the planet marked T, which they push towards Z, it is evident that
they must be diverted by it and constrained to take their course towards
B. I say toward B rather than toward D. For having an inclination to continue their motion in a straight line, they must go toward the outside of
the circle ACZN that they describe, rather than toward the centre S. Now
passing in this way from A to B, they force the planet T to turn with them
about its centre, and conversely this planet, in so turning, gives them 71
occasion to take their course from B to C, then to D and to A, and thus to

45

form about the planet a particular heaven with which it must always afterwards continue to move from the direction that is called 'west' toward that called 'east', not only around the Sun, but also around its own centre.

Moreover, knowing that the planet marked $\mathbb{C}$[75] is disposed to take its course along the circle NACZ, just as is the planet marked T, and that it must move faster because it is smaller, it is readily understood that, wherever in the heavens it might have been located initially, within a short time it had to tend toward the exterior surface of the small heaven ABCD; and once it had joined that heaven, it must subsequently always follow its course around T along with the parts of the second element that are at the surface.

For since we are assuming that it would have exactly as much force as the matter of that heaven to turn along the circle NACZ if the other planet were not there, then we must imagine it to have a little more force to turn along the circle ABCD, because it is smaller and consequently always moves as far away as possible from the centre T. Thus it is that a stone which is moved in a sling always tends to move away from the centre of the circle that it is describing. And yet this planet, being at A, is not thereby going to deviate towards L, since it would then enter a location in the heaven whose matter had the force to push it back towards the circle NACZ. And by the same token, being at C, it is not going to descend toward K, since it would there be surrounded by matter that would provide it with the force to ascend again toward the same circle NACZ. Nor will it go from B toward Z, much less from B toward N, since it could not go there either as easily or as quickly as it could toward C and toward A.[76] So it must remain as if attached to the surface of a small heaven ABCD and turn continually with it about T. That is what prevents its forming another small heaven around it, which would make it turn again around its own centre.

I add nothing here about how a greater number of planets can be found joined together and taking their course about one another, such as those that the new astronomers have observed about Jupiter and Saturn.[77] For I have not undertaken to speak of everything. I have spoken in particular

72

[75] The Moon.
[76] Some kind of principle of least action seems to be being invoked here, although Descartes never explicitly advocates such a principle (as Fermat and Leibniz will).
[77] The reference here is probably above all to Galileo, who first reported the moons of Jupiter in 1610 and those of Saturn in 1611.

about two planets: solely in order to represent to you, by the planet marked T, the Earth we inhabit, and by that marked ☾ the Moon that turns about it.

Chapter 11

On weight[78]

But now I want you to consider what the weight of this Earth is, that is, what the force is that unites all its parts and makes them all tend toward the centre, each more or less according to the extent of its size and solidity. This force is nothing but, and consists in nothing but, the parts of the small heaven which surround it turning much faster than its own parts about its centre, and tending to move away with greater force from its centre, and as a result pushing the parts of the Earth back toward its centre. You may find this presents difficulties, given that I have just said that the most massive and the most solid bodies, such as I have supposed those of comets to be, tend to move outwards to the circumferences of the heavens, and that only those that are less so are pushed back to their centres; as if it followed from this that only the less solid parts of the Earth could be pushed back towards its centre, and that others should move away from it. But note that, when I said that the most massive and solid bodies tended to move away from the centre of a heaven, I was assuming that they were already moving with the same agitation as the matter of that heaven. For it is certain that, if they had not yet begun to move, or if they move at a speed less than that required to follow the course of this matter, they must first be pushed toward the centre around which it is turning; and it is indeed certain that, to the extent to which they are larger and more solid, they will be pushed with more force and speed. Nevertheless, if they are able to compose comets, they will not be prevented from tending to move, a short time later, toward the exterior circumference of the heavens, because the agitation they have acquired in descending toward any one of the heaven's centres will unfailingly provide them with the force to pass beyond it and to ascend again toward its circumference.

In order to understand this more clearly, consider the Earth EFGH [fig. 5], with water 1.2.3.4 and air 5.6.7.8 which, as I shall tell you below,

[73]

[74]

---

[78] The heading in the 1664 edition is: *What weight is.*

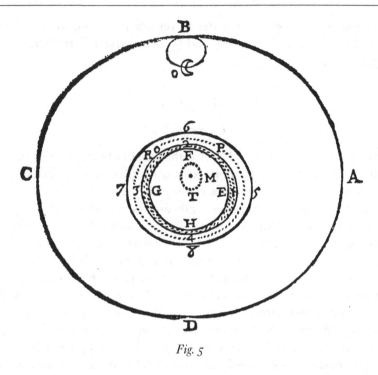

Fig. 5

are composed just of some of the less solid of the Earth's parts, and make up a single mass with it. Next consider also the matter of the heaven, which not only fills all the space between the circles ABCD and 5.6.7.8, but also all the small intervals below it among the parts of the air, the water, and the Earth. And imagine that, as this heaven and this Earth turn together around the centre T, all their parts tend to move away from it, but those of the heaven very much more than those of the Earth, because they are very much more agitated. And we can even imagine that, among the parts of the Earth, those that are more agitated in the same direction as those of the heaven tend to move away from the centre more than do the others. So that, if the entire space beyond the circle ABCD were void, that is, were filled only with matter that was not able to resist the actions of other bodies or to have any significant effect (for this is how we should construe the term 'void'), then all the parts of the heaven in the circle ABCD would leave it first, followed by those of the air and the water, and finally those of the Earth as well, each moving more quickly to the extent that it finds itself less attached to the rest of the mass. In the same way, a stone leaves the sling in which it is being moved as soon as one releases the cord, and the dust one throws on a top while it is turning immediately flies off from it in every direction.

Then consider that, since there is no space such as this beyond the circle ABCD that is void and where the parts of the heavens contained within that circle are able to go, unless others which are exactly similar replace them simultaneously, the parts of the Earth cannot move away any 76 further than they do from the centre T either, unless just as many parts of heaven or other terrestrial parts required to fill them come down to replace them. Nor, conversely, can they move closer to the centre unless just as many others rise in their place. Thus they are all opposed to one another, each one being opposed to those that must replace it should it rise, and to those that must replace it should it descend, just as the two sides of a balance are opposed to one another.[79] That is to say, just as one side of the balance can only be raised or lowered if the other side, at exactly the same instant, moves in the opposite direction, and just as the heavier always makes the other rise, so too the stone R, for example, is so opposed to the quantity of air above it, which is exactly the same size as it and whose place it would have to occupy if it were to move further away from the centre T, that this air would of necessity have to descend to the extent that the stone rose. And in the same way it is opposed to another equal quantity of air below it, whose place it would have to occupy if it were to move closer to that centre T, such that the stone must descend when this air rises.

Now it is evident that, since much more terrestrial matter is contained within this stone than is contained in an amount of air of equal extent, and that to counterbalance this it contains much less celestial matter, and 77 since its terrestrial parts are far less agitated by celestial matter than those of that air, the stone should not have the force to rise above it; but on the contrary this amount of air should rather have the force to make the stone fall downwards. Thus that amount of air is light when compared with the stone, but when compared with the completely pure celestial matter it is heavy. In this way you can see that every part of terrestrial bodies is pressed towards T, not indifferently by all the matter surrounding it but only by an amount of the matter exactly equal to the size of that part and which, being underneath the part, can take its place if that part moves

[79] The image of the balance here is a revealing indication of the source of Descartes' thinking about these mechanical issues: statics.

down. This is the reason why, among the parts of any single body that we call 'homogeneous', such as those of air or water, the pressure on the lowest is not notably more than that on the highest, and why a man at the bottom of very deep water does not feel it pressing on his back any more than if he were swimming right on top.[80]

But if it seems to you that the celestial matter, in making the stone R fall towards T, below the air surrounding it, should also make it travel faster than this air towards 6 or 7 – that is, towards the east or the west – so that the stone does not fall in a straight plumb line as do heavy bodies on the real Earth, consider first that all the terrestrial parts contained in
78 the circle 5.6.7.8, in being pressed toward T by the celestial matter in the way I have just explained, and moreover having very irregular and diverse shapes, must join together and approach one another, and in this way make up only one mass, which is borne as a whole by the course of the heaven ABCD. Thus, while it turns, those of its parts that are at 6, for example, always remain opposite those that are at 2 and at F, without any appreciable movement to one side or the other except in so far as winds or other particular causes make them do so.

And note also that the little heaven ABCD turns much faster than this Earth, but those of its parts that are caught in the pores of terrestrial bodies cannot turn appreciably faster than these bodies about the centre T, even though they move much faster in many other directions, depending on the disposition of those pores.

Next you should know that even though the celestial matter makes the stone R approach this centre, because it tends to move away from it with more force than the stone, it cannot for all that force the stone back up towards the west, despite the fact that it tends to move towards the west with a greater force than the stone. For consider that this celestial matter tends to move away from the centre T because it tends to continue its motion in a straight line; nevertheless, it tends to move from west to east only because it tends to continue in its motion at the same speed and, moreover, because it does not make the slightest difference to it whether
79 it finds itself at 6 or at 7.

Now it is evident that its motion is slightly more rectilinear while it is causing the stone R to fall towards T than it is in leaving the stone at R; but if it caused the stone to move back towards the west it would not be able

---

[80] Descartes' ignorance of the basic principles of hydrostatics here is somewhat surprising.

to move as quickly towards the east as it would if it left the stone where it was, or even if it pushed the stone in front of it.

But you should know that even though this celestial matter has a greater force to cause the stone R to descend towards T than it has to cause the air surrounding the stone to descend there, it should not have a greater force to push it in front of it from west to east, nor consequently to cause the stone to move faster in that direction than it does the air. For consider that there is exactly the same amount of this celestial matter acting on the stone to cause it to descend towards T, and that it uses its full force to that end, as there is terrestrial matter in the composition of the stone's body; also that, inasmuch as there is much more terrestrial matter in the stone than there is in the same amount of air, the stone must be pressed much more strongly than the air in the direction of T. But making the stone turn towards the east, all the celestial matter contained in the circle R acts on it, as well as on the terrestrial parts of the air contained in this circle. Thus, there being no more action against it than against the air, the stone should not turn faster than the air in that direction. 80

And you will understand from this that the arguments that a number of Philosophers employ to refute the motion of the actual Earth have no force against the motion of the Earth that I am describing to you. When they say, for example, that if the Earth moved heavy bodies could not descend in a straight line towards the centre, but would move, rather, here and there in the direction of the heavens; and that cannons pointed towards the west should have a greater range than if pointed towards the east; and that one should always feel great winds in the air and hear great noises; these and similar things occur only on the assumption that the Earth is not carried by the course of the heaven surrounding it, but that it is moved by some other force and in some other direction than that heaven.[81]

[81] Note how different Descartes' response to these standard objections to the Earth's rotational motion is from that of Galileo in his *Two Chief World Systems*. Galileo invokes a kinematic principle of relativity of motion to account for the facts that cannon balls fired to the east and to the west have the same range and that bodies fall vertically downwards. On Galileo's principle, motion can be resolved into various components, and we are only aware of those in which we do not share. So, in the case of the fall of bodies, these actually describe a parabolic path, and this is what we would see if we were to view the motion from an inertial frame stationary with respect to the Earth's rotation; but because we share in the rotational motion, we are not aware of this component, only of the downward motion, and this is what we actually see. Descartes' account appeals to no such kinematic principle: rather, he denies that the Earth is moving in a medium (the surrounding air) which is stationary with respect to it, and asserts that the Earth carries the medium around with it. The air through which the body falls is really moving as much as the Earth is.

Chapter 12

On The ebb and flow of the sea

Having thus explained the weight of the parts of this Earth, which is caused by the motion of the celestial matter in their pores, I must now discuss a particular motion of its whole mass, which is caused by the presence of the Moon, as well as some peculiarities that depend on that motion.

To this end, consider the Moon to be at B for example [see fig. 5], where you can assume that it is stationary in comparison to the speed at which the celestial matter below it moves. Consider also that this celestial matter, having less space to traverse between o and 6 than between B and 6 (if the Moon does not occupy the space between o and B), and consequently having to move a little more quickly there, cannot but have the force to push the whole Earth slightly towards D, so that, as you can see, its centre T moves away slightly from the point M, which is the centre of the small heaven ABCD. For the course of the celestial matter is all that keeps the Earth where it is located. And because the air 5.6.7.8 and the water 1.2.3.4 surrounding this Earth are fluid bodies, it is evident that the same force that presses against the Earth in this way must also make them sink towards T, not only from the side 6.2 but also from the opposite side 8.4; and in compensation must make them rise at 5.1 and 7.3. In this way, since the surface EFGH of the Earth remains round, because it is hard, that of the water 1.2.3.4 and the air 5.6.7.8, which are fluid, must take the shape of an oval.[82]

Next, consider that since the Earth is turning around its centre in the meantime, and by this means producing the days, which we divide up into 24 hours, as we do ours, the side F, which is now facing the Moon, and on which, because of this, the water 2 is not as high, must be facing the heaven marked C in six hours, where this water will be higher, and in twelve hours will be facing the spot in the heavens marked D, where the water will again be lower. Thus the sea, which is represented by this water 1.2.3.4, must have its ebb and flow around this Earth once every six hours, just as it has around that in which we live.

---

[82] It is possible that part of the reason why Descartes went to Sweden at the end of 1649 was to show that, because of the oval shape of the atmosphere, as one travelled further north there should be a detectable difference in barometric pressure. Mersenne and the French ambassador to Sweden, Chanut, had urged him to make such experiments.

Consider also that while this Earth turns from E, through F, to G – that is, from the west through the meridian to the east – the swell of water and air that remains at 1 and 5, and at 3 and 7, passes from its eastern to its western side, giving rise there to an ebb without flow very similar to that which, according to the reports of our pilots, makes navigation on our seas from east to west very much easier than that from west to east.

And so that nothing is overlooked at this point,[83] let us add that the Moon makes the same circuit in a month as the Earth does each day, so causing the points 1.2.3.4 that mark high and low water to advance gradually towards the east. As a consequence, these waters do not change precisely every six hours, but rather lag behind by about a fifth part of an hour each time, just as those of our seas do.

Consider in addition that the small heaven ABCD is not exactly round, but that it is a little more freely extended at A and C, and moves proportionately more slowly there than at B and at D, where it cannot interrupt the course of the matter of the other heaven enclosing it so easily. Thus the Moon, which always remains as if it were attached to its exterior surface, has to move a little more quickly and vary a little less in its path; and consequently this is the reason why the ebb and flow of the sea are much greater when the Moon is at B, where it is full, and at D, where it is new, than when it is at A or at C, where it is only half. These are peculiarities which are also just like those that astronomers observe in the actual Moon, although they are perhaps unable to explain them as easily on the basis of the hypotheses that they use.

As for the other effects of this Moon, which differ depending on whether it is full or new, they manifestly depend on its light. And as far as the other peculiarities of the ebb and flow of the sea are concerned, they depend in part on the different coastal situations, and in part on the prevailing winds at the time and place of observation. Finally, as for the other general motions, of the Earth and the Moon as well as of the stars and the heavens, either you will understand them sufficiently from what I have said, or they do not come in my purview; not coming under the same project as those of which I have spoken, they would take too long for me to describe. Thus all that remains for me to do here is to explain this action of the heavens that I said earlier should be taken to be their light.

[83] Actually, Descartes does omit one thing here, the half-yearly tidal cycle. This will be dealt with in the more complete account of the *Principles*.

84

Chapter 13

On light[84]

I have already said on a number of occasions that revolving bodies always tend to move away from the centres of the circles that they describe. But here I must determine in which direction the parts of matter that the heavens and stars are composed of tend.

To this end, it should be noted that when I say that a body tends in some direction, I do not thereby want anyone to imagine that there is a thought or will in the body that bears it there, but only that it is disposed to move there, whether it actually moves or whether some other body prevents it from doing so.[85] And it is principally in this last sense that I use the word 'tend', because it seems to signify some exertion and because every exertion presupposes some resistance. Now in so far as there are often a number of different causes which, acting together on the same body, impede one another's effect, one can, depending on various considerations, say that the same body tends in different directions at the same time. We have just said that the parts of the Earth tend to move away from its centre to the extent that they are considered in isolation, but that on the other hand they tend to move closer to it to the extent that one considers the parts of the heaven pushing them there, and again that

85 they tend to move away from it if they are considered as opposed to other terrestrial parts that compose more massive bodies.

Thus the stone turning in a sling along the circle AB, for example [see fig. 1], tends towards c when it is at point A, if one considers just its agitation in isolation; and it tends circularly from A to B, if one considers its motion as regulated and determined by the length of the cord which retains it; and finally the same stone tends towards E if, ignoring that part of its agitation whose effect is not impeded, the other part of it is opposed to the resistance that this sling continually offers to it.

But for a distinct understanding of this last point, imagine this stone's inclination to move from A to c as if it were composed of two other

---

84 The heading in the 1664 edition is: *What light consists in.*

85 In other words, tendencies to motion do not have to be realised in the form of actual motions, for the body may have a tendency to move in a particular direction but may be prevented (e.g. by a force) from so moving. It is central to Descartes' account that it is tendencies rather than actual motions that are at stake. For most purposes, tendencies to motion are effectively the orthogonal components of motion into which Descartes resolves motions.

inclinations, one turning along the circle AB and the other rising straight up along the line VXY. And imagine the proportion of the inclinations were such that if the stone were at the position on the sling marked V when the sling was at the position on the circle marked A, it should subsequently be at the position marked X when the sling is at B, and at the position marked Y when the sling is at F, and thus should always remain in the straight line ACG. Then, since we know that one of the parts of its inclination, namely that which carries it along the circle AB, is in no way impeded by this sling,[86] it is easily seen that the stone meets resistance only in its other part, namely that which would cause it to move along the line DVXY if it were unimpeded, and consequently it tends – that is, strives – only to move away from the centre D. And note that, considered in this way, when the stone is at point A it tends so exactly toward E that it is not at all more disposed to move toward H than toward I, although it would be easy to persuade yourself of the contrary if you failed to consider the differences between the motion that it already has and the inclination to move that remains with it.

86

Now each of the parts of the second element that compose the heavens should be thought of as being the same as this stone; that is, those that are at E [in fig. 5], say, tend of their own inclination toward P, but the resistance of the other parts of the heaven which are above them cause them to tend – that is, dispose them to move – along the circle ER. This resistance in turn is opposed to the inclination that they have to continue their motion in a straight line – that is, is the reason why they strive to move – toward M. And so, accounting for the others in the same way, you can see in what sense one can say that they tend toward the places that are directly opposite the centre of the heaven that they compose.

87

But there is more to be considered in the parts of the heaven than in a stone turning in a sling. The parts are continually pushed, both by all those similar parts between them and the star that lies at the centre of their heaven, and by the matter of that star; and they are not pushed at all by the others. Those at E, for example [fig. 6], are not pushed by those at M, at T, at R, at K, or at H, but only by all those between the two lines AF and DG, together with the matter of the Sun. This is the reason why they

88

---

[86] This seems to be a statement of circular inertia. It is possible that, because his model is taken from statics, he thinks of circular motion in terms of a state of equilibrium, and somehow equates equilibrium and inertia.

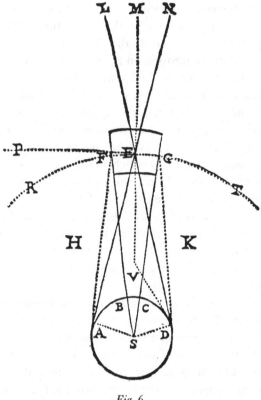

Fig. 6

tend not only toward M, but also toward L, toward N, and generally toward all the points that can be reached by all the rays or straight lines which come from some part of the Sun and pass through the place where the parts are.[87]

But so that the explanation of all this be easier, I want you to consider the parts of the second element alone, as if all the spaces occupied by the matter of the first element – both where the Sun is and elsewhere – were void. Indeed, because there is no better means of knowing whether a body is pushed by others than to see if these others would advance towards the place where it is, in order to fill that place if it were to become empty, I also want you to imagine that the parts of the second element at E are

---

[87] As Mahoney, *René Descartes, Le Monde*, remarks in his notes (pp. 221–2 n. 58), E is both a point which is the apex of the visual cone EAD and a space around this point.

removed from it. And, having assumed this, note in the first place that none of those above the circle TER, such as at M, are at all disposed to fill their place, more especially as, on the contrary, each of these tends to move away from it. Then note also that those in that circle, namely at T, are no more disposed to do so, for even though they do indeed move from T toward G along the course of the whole heaven, nevertheless, because those at F also move with the same speed toward R, the space E, which must be imagined to be capable of moving like them, cannot but remain 89 void between G and F, provided others do not come from elsewhere to fill it. And in the third place, those that are below that circle but not contained between the lines AF and DG, such as those at H and at K, also do not tend at all to advance toward space E to fill it, even though the inclination they have to move away from point S so disposes them in some way, just as the weight of a stone disposes it not only to descend along a straight line in the free air, but also to roll down a mountain sideways in those circumstances where it cannot descend any other way.

Now what stops them from tending toward that space is the continuation of all motions, in so far as is possible, in a straight line, and consequently when Nature has many ways of arriving at the same effect, she always unfailingly follows the shortest. For if the parts of the second element which are, say, at K, advanced toward E, all those that were closer to the Sun than they would also advance at the same instant toward the place they were vacating, and so the effect of their movement would just be that the space E would be filled and another of equal size in the circumference ABCD would become void at the same time. But the same effect can clearly follow much better if those that are between the lines AF 90 and DG advance immediately toward E, with the result that when there is nothing preventing these parts from doing this, the others do not tend at all toward E, any more than a stone tends to fall obliquely toward the centre of the Earth when it can fall in a straight line.

Finally, consider that each of the parts of the second element between the lines AF and DG must advance together toward this space E in order to fill it at the instant it becomes void. For even though what carries them toward E is only the inclination that they have to move away from point S, and this inclination causes those that are between the lines BF and CG to tend more directly toward E than those that remain between the lines AF and BF, and DG and CG, you will see nevertheless that these latter parts are just as disposed as the others to go there, if you note the effect that must

follow from their motion which, as I have just said, is simply that space E is filled, and that there is another of equal size in the circumference ABCD that becomes void at the same time. For the change of position which they undergo in the other places that they previously filled, and which remain full of them afterwards, is not at all significant, in that they are so completely alike that it makes no difference which parts of matter fill these
91 places. Nevertheless, note that one must not conclude from this that they are all equal, merely that the motions that can cause them to be unequal are irrelevant to the action we are dealing with.

Now there is no shorter means of filling one part E of space while another, for example at D, becomes empty, than for all the parts of matter on the straight line DG, or DE, to advance together toward E. For if it were only those between the lines BF and CG that were to advance first towards this space E they would leave another space beneath them at V, into which those that were at D would have to come. In this way the same effect that the motion of matter in the straight line DG or DE can produce would be made by the motion of that in the curved line DVE. And this is contrary to the laws of motion.

But you may find it a little difficult here to understand how the parts of the second element between the lines AF and DG can go forward together toward E since, because the distance between A and D is greater than that between F and G, the space they must enter in order to go forward in this way is narrower than that which they must leave. Bear in mind that the action by which they tend to move away from the centre of the heaven does not force them to touch those of their neighbours that are
92 at the same distance as they are from that centre, but rather those that are slightly more distant from it, in the same way as the weight of the small balls 1, 2, 3, 4 [fig. 7] does not make those with the same numeral touch one another, but only makes those marked 1 or 10 rest on those marked 2 or 20, and these to rest on those marked 3 or 30, and so on. Thus, these small balls can be arranged not just as in figure 7, but as they are in figures 8 and 9, and in a thousand other different ways.

Next consider that parts of the second element, which move – as we
93 said above they must – separately from one another, cannot ever be arranged like the balls in figure 7. But it is only in this way that the difficulty postulated can arise. For one cannot suppose there to be, between those parts of the heaven that are the same distance from the centre of their heaven, an interval which is so small that it would prevent

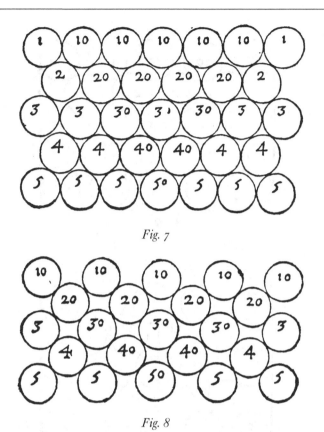

Fig. 7

Fig. 8

us imagining the inclination that they have to move away from that centre forcing those between the lines AF and DG to go forward all together toward the space E when it is empty. Thus if you compare figure 9 with figure 10, you will see that the weight of the small balls 40, 30, etc. must cause them to descend all together toward the space occupied by that ₉₄ marked 50 as soon as this one can vacate it.

And one can see clearly that here how those of the balls that are marked with the same numeral are arranged in a space which is narrower than that which they leave, namely by moving closer to one another. One can also see that the two balls marked 40 must descend a little faster and move proportionately closer to one another than the three marked 30, and these three must move faster and closer to one another than the four marked 20, and so on.

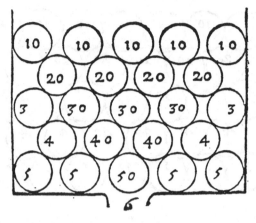

Fig. 9

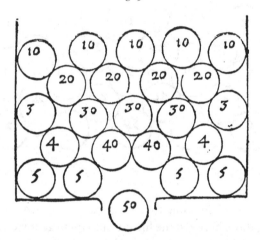

Fig. 10

At this point you will perhaps say to me that it appears from figure 10 that the two balls numbered 40 and 40, after having descended a little, come to touch one another, which is why they stop without being able to descend further. In exactly the same way, the parts of the second element 95 that must advance toward E will stop before having completely filled the whole space we have assumed to be there.

But I reply to this that their being able to advance toward E at all is sufficient to establish perfectly what I have said, namely that since the whole space that is there is already filled by some body (whatever it might be), the parts press continually on that body and strive against it as if to chase it out of its place.

And furthermore, I reply that, since their other motions, which continue in them while they are advancing in this way toward E, do not allow them to remain arranged in the same way for one moment, they prevent them from touching one another; or, rather, cause them upon touching one another to separate again immediately, and so there is nothing here that prevents them advancing without interruption toward the space E, until it is completely filled. Thus all that can be concluded from this is that the force with which they tend toward E is perhaps as if it were trembling, increasing and relaxing in various small tremors, as the parts change position, a property which seems to be rather suited to light.

Now if, on the assumption that the spaces E and S and all the small angles between the parts of the heaven are empty, you have understood this adequately, you will understand it even better by supposing them to be filled with the matter of the first element. For the parts of this element found in the space E cannot prevent those of the second element between the lines AF and DG advancing to fill it up, in just the same way as they would if it were a void, because, being extremely subtle and extremely agitated, they are as ready to leave the places they occupy as other bodies are to enter them. And for this reason, those that occupy the small angles between the parts of the heaven give up their place without resistance to those that come from that space E and are in the direction of point S. I say S rather than any other place because the other bodies, which are more unified and larger and so have more force, tend to move away from it.

And it should be noted that they pass from E toward S between the parts of the second element that go from S toward E, without the one in any way impeding the other. Thus the air enclosed in the hourglass XYZ [fig. 11] rises from Z toward X through the sand Y, which however still falls in the meatime toward Z.

Finally, the parts of that first element that occupy the space ABCD, where they compose the body of the Sun, turn around the point S very rapidly, tending to distance themselves from it in all directions in a straight line, in accordance with what I have just set out; and by these means, all those that are in the line SD jointly push the part of the second element which is at point D, and all those in line SA push that which is at

Fig. 11

point A, and so on. And they do this in such a way that it is sufficient by itself to make all those parts of the second element that are between the lines AF and DG advance toward the space E, even though in themselves they might have no inclination to do so.

Moreover, since they have to advance in this way toward this space E, when it is occupied just by the matter of the first element, it is certain that they also tend to go there even when it is full of some other kind of body and consequently that they push and strive against that body to drive it out of its place. And so, if it were the eye of a man that was at point E, it would really be pushed, both by the Sun and by all the celestial matter between the lines AF and DG.

Now one must know that the men of this new world will be of such a nature that, when their eyes are pushed in this fashion, they have a sensation very similar to that which we have of light, as I shall explain more fully below.

Chapter 14

On the properties of light

But I want to pause a while at this point to explain the properties of the action by which our eyes can be thus pushed. For these are all in such
98 perfect accord with those we note in light that, when you have considered them, I am sure you will allow, as I do, that there is no need to imagine there to be in the stars or in the heavens any quality other than this action that is called by the name of 'light.'

The principal properties of light are: (1) that it extends circularly in all directions around those bodies one calls luminous; (2) to any distance whatever; (3) instantaneously; (4) and ordinarily in straight lines,[88] which should be taken as rays of light; (5) and that several of these rays coming from different points can collect together at the same point; (6), or, coming from the same point, can go out toward different points; (7) or, coming from different points and going to different points, can pass through the same point without impeding one another; (8) and that they can sometimes impede one another, namely when they are of very unequal force, that of some rays being far greater than that of others; (9) and, finally, that they can be diverted by reflection; (10) or by refraction; (11) and that their force can be increased, (12) or diminished by the different dispositions or qualities of the matter that receives them. Here are the principal qualities observed in light, and all of them are in accord with this action, as you shall see.

(1) The reason why this action should extend in all directions around luminous bodies is evident: it is that it proceeds from the circular motion of their parts. 99

(2) It is also evident that it can extend to any distance. For if we suppose, for example, that the parts of the heaven located between AF and DG are already themselves disposed to advance toward E, as we have said they are, it can no longer be doubted that the force with which the Sun pushes those at ABCD should also extend out to E, even if the distance between the two were greater than that between the highest stars of the firmament and us.

(3) And knowing that the parts of the second element between AF and DG all touch and press one another as much as possible, one cannot doubt either that the action by which the first ones are pushed must instantaneously pass through to the last, in just the same way that the force by which one end of a stick is pushed passes through to the other end in the same instant.[89] Or rather – in case you will object on the grounds that parts of the heavens are not attached to one another in the way that those of a stick are – just as the small ball marked 50 [in fig. 9] falls toward 6, the others marked 10 also fall toward 6 in the same instant.

(4) As for the lines along which this action is communicated, and which 100

---

[88] The qualification, 'ordinarily' in straight lines, is presumably meant to indicate that refraction is not being taken into account.

[89] This image, a favourite one of Descartes', would have been accepted by all his contemporaries, even those who believed that the speed of light must be finite. Only with the advent of the Special Theory of Relativity was it shown that there is a natural limit to the speed of propagation of a physical action.

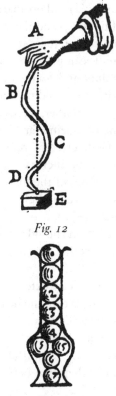

Fig. 12

Fig. 13

are in fact light rays, it must be noted that they differ from the parts of
the second element by means of which this action is communicated; and
they are not something material in the medium through which they pass,
but merely indicate in what direction and with what determination[90] the
luminous body acts on the body it is illuminating. Thus we should not
stop thinking of them as perfectly straight even though the parts of the
second element that serve to transmit this action – that is, light – can
almost never be placed so directly one on the other that they make exactly
straight lines. In just the same way, we can readily conceive that a hand A
[fig. 12] pushes the body E along the straight line AE even though it only
pushes it through the intermediary of the stick BCD, which is twisted. And
in just the same way, you can think of the ball marked 1 [fig. 13] pushing

---

90 On the term 'détermination' here, see the Introduction.

Fig. 14

that marked 7 through the intermediary of the two marked 5 and 5 as directly as through the intermediary of the others, 2, 3 , 4, and 6.

(5) (6) It is also easy to understand how several of these rays, coming from different points, meet at the same point, or coming from the same point, go out toward different points, without impeding or relying on one another. As you can see from Fig. 6, several of them coming from the points A, B, C, and D come together at point E; and several come down from the single point D and extend, one toward E and one toward K, and thus towards innumerable other places. In just the same way, the various forces with which the cords 1, 2, 3, 4, and 5 [fig. 14] are pulled all come together in the pulley, and the resistance of the pulley extends to all the different hands that are pulling those cords.

(7) But to understand how several of those rays, coming from and going toward various points, can pass through the same point without impeding one another – just as in fig. 6 the two rays AN and DL pass through point E – one must regard each of the parts of the second element as being able to receive several different motions at the same time. Thus the part at point E, for example, can be pushed as a whole toward L by the action coming from the place on the Sun marked D, and simultaneously toward N by that coming from the place marked A. You will understand this better if you consider that air can be pushed from F toward G, from H toward I, and from K toward L [fig. 15] all at the same time, through the three tubes FG, HI, and KL, even if the tubes are joined at point N in such a way that all the air that passes through the middle of one of them must necessarily also pass through the middle of the other two.

(8) And this same comparison can serve to explain how a strong light prevents the effect of those that are weaker. For if we push the air through

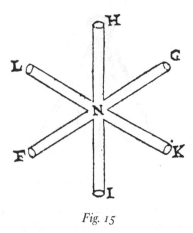

Fig. 15

F much more strongly than through H or through K, it will not tend at all toward I or toward L, but only toward G.

(9) (10) As for reflection and refraction, I have already explained them sufficiently elsewhere.[91] Nevertheless, because I used the example of the motion of a ball there instead of speaking of light rays, so as to make my account more easily understood, it still remains for me here to have you consider that the action, or the inclination to move, that is transmitted from one place to another through several bodies which touch one another and continuously fill the whole space between them follows precisely the same path along which this same action would cause the first of these bodies to move were the others not in its way. The sole difference is that it requires time for that body to move whereas the action in it can extend to any distance instantaneously by means of those touching it. It follows from this that, just as a ball is reflected when it strikes against the wall of a tennis court, and is refracted when it enters or leaves a body of water obliquely, so too when the light rays meet a body that does not allow them to pass beyond it, they must be reflected; and when they enter obliquely some place through which they can spread more or less easily than they are able to in that through which they are coming, they must also be deflected and undergo refraction at the point of that change.

(11) (12) Finally, the force of light is not only greater or smaller in each place depending on the quantity of rays that come together there, but it can also be increased or diminished by the various dispositions of the

103

[91] See translation of Discourse 2 of the *Dioptrics*, below.

bodies in the places through which it passes. In the same way, the speed of a ball or a stone which is pushed into the air can be increased by winds blowing in the same direction as its motion, and diminished by their contraries.

Chapter 15

104

That the face of the heaven of this new world must appear to its inhabitants completely like that of our world[92]

Having thus explained the nature and properties of the action I have taken to be light, I also need to explain how, by its means, the inhabitants of the planet I have assumed to be like the Earth can see the face of their heaven to be just like that of ours.

First, there is no doubt at all that they must see the body marked s [see fig. 4] as completely full of light, and similar to our Sun, given that this body sends light rays from all points of its surface toward their eyes. And because it is much closer to them than the stars, it must appear to them very much larger. It is true that the parts of the small heaven ABCD which turns around the Earth offer some resistance to those rays; but because all those of the large heaven that are between s and D strengthen the rays, those between D and T, being comparatively few in number, can remove very little of their force from them. And even all the action of the parts of the large heaven FGGF [see fig. 2] is not enough to prevent them reaching as far as the Earth from the side on which it is not illuminated by the Sun.

For it must be understood that although the large heavens – that is, those that have a fixed star or the Sun for their centre – may perhaps be quite unequal in size, they must always be of exactly equal force. Because of this, all the matter in the line SB, for example, must tend as strongly toward ϵ as that which is in the line ϵB tends toward s.

Now since the whole force of the ray SB, for example, is just exactly equal to that of the ray ϵB, that of the ray TB, which is less, manifestly cannot prevent the force of the ray ϵB extending to T. And in the same way it is evident that the star A can extend its rays to the Earth T, to the extent that the matter of the heaven between A and 2 helps them more than that between 4 and T resists them, and also to the extent that that between 3 and 4 helps them no less than that between 3 and 2 resists them. And so,

105

---

[92] The heading in the 1664 edition is: *The Way in which the Sun and the Stars act on our Eyes.*

judging the others in the same proportions, you can understand that these stars cannot appear arranged in a less confused way, or fewer in number, or with fewer inequalities between them, that do those we see in the actual world.

But as regards their arrangement, you must understand nonetheless that they can hardly ever appear to be in their actual places. That marked ε, for example, appears as if it were along the straight line TB, and the other marked A as if it were along the straight line T4. This is because, the heavens being unequal in size, the surfaces that separate them are hardly ever so disposed that the rays that pass through them going from the stars towards the Earth meet them at right angles. And when they meet them obliquely, it follows with certainty from what has been demonstrated in the *Dioptrics*,[93] that they must bend and undergo significant refraction, to the extent that they pass through one side of this surface much more easily than through the other. And these lines TB, T4, and the like must be supposed to be so extremely long in comparison to the diameter of the circle that the Earth describes around the Sun that, wherever the Earth lies on this circle, the people on it will always see the stars as fixed and attached to the same places on the firmament, or, to use the astronomers' terms, they cannot observe parallax in stars.[94]

As regards the number of these stars, consider also that the same star can often appear in different places due to the different surfaces that deflect its rays toward the Earth. That marked A [in fig. 2], for example, appears both along the line T4, by means of the ray A24T, and along the line Tf, by means of the ray A6fT, in the same way as objects are multiplied when viewed through glasses or other transparent bodies with many-faceted surfaces.

What is more, as regards their size, consider that they must appear very much smaller that they are, because of their extreme distance. And for this reason, most of them cannot appear at all, and others only in so far as the rays of several joined together make those parts of the firmament through which they pass a little whiter, and similar to certain stars the

106

107

---

[93] Discourse 1: AT vi. 81–93.

[94] This is a key point in the defence of Copernicanism, since for any distance of the order of magnitude of the Earth's distance from the Sun, we would naturally expect to note differences in the positions of fixed stars depending on whether we viewed them from one extreme of the Earth's orbit or the other. The only way to explain the lack of parallax is to make the distance between the Earth and the fixed stars so immense that the difference between the extremes of the Earth's orbit would be insignificant.

astronomers call 'nebulous', or to that great belt of our heaven that the poets imagine to be whitened by the milk of Juno.[95] But despite this, it is enough to suppose the less distant stars to be roughly equal to our Sun for us to judge that they can appear as large as the largest of our world.

Generally speaking, bodies that send out stronger rays against the eyes of onlookers than do those surrounding them appear proportionately larger than they, and consequently these stars must always appear larger than the parts of the heavens that are equal and adjacent to them, as I will explain below. And as well as this, the surfaces FG, GG, GF and ones like them, where the refractions of their rays occur, can be curved in such a way that they increase their size very significantly; and indeed they increase it even when they are flat. 108

Moreover it is highly likely that these surfaces, being very fluid and constantly moving, should always shake and quiver slightly, and consequently that the stars one sees through them should appear to scintillate and tremble, as it were, just as ours do, and even, because of this trembling, appear slightly larger.[96] Thus the image of the Moon appears larger when viewed from the bottom of a lake whose surface is not very disturbed or agitated, but just slightly rippled by some breath of wind.

And finally it can happen that in the course of time those surfaces change slightly, and indeed some of them bend quite noticeably in a short time, even if only when approached by a comet. In this way, after a long time, several stars seem to change position slightly without changing size, or change size slightly without changing position. Indeed, some even begin to appear or disappear quite suddenly, just as has been observed in the actual world.[97]

As regards the planets and the comets that are in the same heaven as the Sun, since we know that the parts of the third element that compose them are so large or so joined together from several pieces that they can resist the action of light, it is easy to understand that they should appear by means of the rays that the Sun sends toward them, and which are reflected from there toward the Earth, just as the opaque or dark objects 109

[95] In his *Sidereus nuncius* (1610), Galileo had reported that the Milky Way was made up of 'congeries of innumerable stars grouped together in clusters', thereby effectively settling a question that had been disputed since antiquity.

[96] Descartes seems to be concerned with explaining the twinkling or 'irradiation effect' of stars here.

[97] The reference here is to the new stars or *novae* that had been observed, one close to the edge of the Milky Way in 1572, and the other in Ophiuchus in 1604. The discoveries were of importance in demonstrating the falsehood of the Aristotelian doctrine of immutability in the celestial sphere.

that are in a room can be seen there by means of the rays that the lamp shining there sends toward them and that return from them to the eyes of onlookers. And the Sun's rays have, moreover, a very distinct advantage over those of a lamp. This lies in their forces being conserved, or even being augmented increasingly to the extent that they move away from the Sun, until they have reached the outer surface of its heaven, because all the matter of that heaven tends toward there. The rays of a lamp, by contrast, are weakened the further away they are, in proportion to the size of the spherical surface they illuminate; and still more because of the resistance of the air through which they travel. And so the objects close to the lamp receive noticeably more illumination from it than those far away; whereas the lowest planets[98] do not receive proportionately more illumination from the Sun than the highest, nor even more than the comets, which are incomparably more distant.

Now experience shows us that the same thing happens in the actual world as well. I do not believe, however, that it is possible to account for this if one supposes light to be anything other than an action or disposition in objects such as I have explained. I say an action or disposition, for if you have attended carefully to what I have just shown – namely that, if the space where the Sun is were completely empty, the parts of its heavens would tend constantly toward the eyes of onlookers in the same way as when they are pushed by its matter, and even with almost as much force – you can well judge that there is almost no need to have any action at all in the Sun itself, nor even for it to be anything other than pure space in order for it to appear as we see it. This is something that earlier, perhaps, you would have taken to be something very paradoxical. Moreover, the motion of planets around their centres is what makes them twinkle, though much less strongly and in a different way from the fixed stars. And because the Moon lacks this motion, it does not twinkle at all.

As for the comets that are not in the same heaven as the Sun, they cannot send out anything like as many rays toward the Earth as they could were they in the same heaven, not even when they are on the verge of entering it. Consequently, they cannot be seen, except perhaps when they are of extraordinary size. This is because most of the rays that the Sun sends out toward them are carried here and there and dissipated as it were by the refraction that they undergo in the part of the firmament through

[98] That is, those closest to the Sun, namely Mercury and Venus.

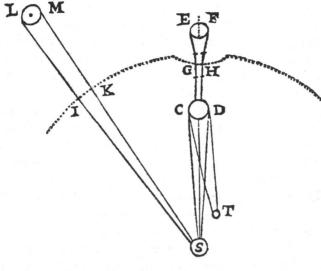

Fig. 16

which they pass. Whereas the comet CD, for example [fig. 16], receives
from the Sun, marked S, all the rays between the lines SC and SD, and sends   111
back toward the Earth all those between the lines CT and DT,[99] one must
think of the comet EF receiving from the same Sun only the rays between
the lines SGE and SHF because, passing much more easily from S to the
surface GH, which I take to be a part of the firmament that they cannot
pass beyond, their refraction there must be very great and very much outward. This diverts many of them to go in the direction of the comet EF.
Note above all that this surface is curved inward toward the Sun, as you
know it should curve when the comet approaches it. But even if it were
completely flat, or even curved in the other direction, the majority of the
rays that the Sun sent out to it could not fail to be prevented by the refrac-   112
tion, if not from going as far as it, at least from returning from there to
the Earth. If, for example, one supposes the part of the firmament IK to
be a portion of a sphere with its centre at S, the rays SIL and SKM should
not bend there at all in going toward the comet LM. But by the same token
they should bend very considerably in returning from the comet toward
the Earth, so that they can reach it only very enfeebled and in very small
quantity. Beyond that, since this can happen when the comet is still very

---

[99] As Mahoney (op. cit., p. 223 n. 69) points out, Descartes' mathematics is extraordinarily sloppy
here: C and D cannot be common points of tangency unless T coincides with S.

far from the heaven containing the Sun (for otherwise, if it were close to it, it would cause its surface to curve inwards) its distance also prevents it from receiving as many rays as when it is ready to enter the heaven. As for the rays it receives from the fixed star at the centre of the heaven that contains it, it cannot send them towards the Earth any more than the new Moon can send back those of the Sun.

But what is more remarkable about these comets is a particular refraction of their rays, which is what usually causes some of them to appear in the form of a tail or curl[100] around them. You will understand this easily if you cast your eyes on [fig. 17], where s is the Sun, c a comet, EBG the

113 sphere which, following what we have said above, is composed of those parts of the second element that are the largest and least agitated of all, and DA the circle described by the annual motion of the Earth. And then

114 imagine that the ray coming from c towards B passes directly to point A, but also that at point B it begins to grow larger and is divided into many other rays which extend here and there in all directions, in such a way that each of them finds itself that much weaker the further it is carried away from that in the middle, BA, which is the principal ray and the strongest one. And then, when the ray CE is at point E, it begins to grow larger and divides into many other such as EH, EY, ES, the principal and strongest of these being EH, the weakest being ES. And similarly, CG passes mainly from G toward I, but as well as this it is also carried away from s and toward all the spaces between GI and GS. Finally, all the other rays that one can imagine between these three – CE, CB, and CG – more or less follow the behaviour [*nature*] of each of these, depending on how close they are. To this I might add that they should be bent slightly toward the Sun; but this is not at all necessary for my purposes, and I often leave out many things so as to make those I do explain all the more simple and easy.

Now given this refraction, when the Earth is at A, it is clear that not only should the men on it be caused to see the body of the comet c by the ray BA; but also that the rays LA, KA, and similar ones, which are weaker, should give the appearance to their eyes of a corona or tail of light spread

115 out uniformly in all directions around the comet (as you see at the place marked 11), at least if they are strong enough to be perceived, which they

---

100 The term that Descartes uses here – *chevelure* – literally means a 'head of hair'. In modern French it is the term for the tail of a comet, but here and below Descartes distinguishes between a tail (*queuë*) and a *chevelure*, telling us below that it is transformed from a *chevelure* into a *queuë*. Since the distinction is a real one, I have followed Mahoney, op. cit., in translating the term *chevelure* as 'curl'.

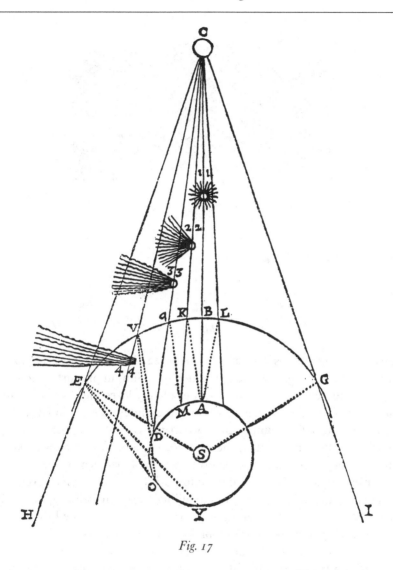

Fig. 17

can often be if they come from comets – but not if they come from planets, or even from fixed stars, which should be imagined to be smaller.

It is also clear that when the Earth is at M and the comet appears by means of the ray CKM, its tail should appear by means of QM and all the other rays that tend toward M, so that it extends further than before toward the part opposite to the Sun, and less or not at all toward the

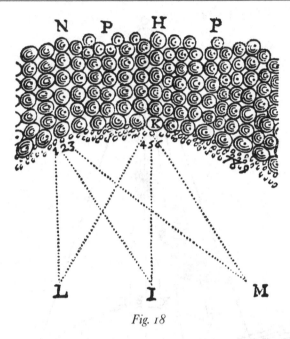

Fig. 18

person looking at it, as you can see here at 22. And thus appearing to be
longer and longer on the side opposite to the Sun, in proportion to how
far away the Earth is from point A, it gradually loses the shape of a curl
and is transformed into a long tail, which the comet trails behind it. When
the Earth is at D, for example, the rays QD and VD make it appear as at 33.
And when the earth is at O, the rays VO, EO, and similar ones make it
appear to be still longer. And finally, when the Earth is at Y, one can no
longer see the comet because of the interposition of the Sun; however, the
rays VY, EY, and other similar ones unfailingly cause its tail to appear still
as a Λ-shape or a curl, as at 44 here. And it should be noted that since
neither the sphere EBG, nor any of those it contains, is always exactly
116 round, as can easily be gleaned from our account, these tails or torches
need not always appear straight, nor entirely in the same plane as the Sun.

As for the refraction that is responsible for all of this, I confess that its
nature is very special and very different from that commonly observed
elsewhere. But you will not fail to see clearly that it must take place in
the way I have just described to you when you consider that the ball H
[fig. 18] , being pushed toward I, also pushes all those beneath it as far
down as K toward I, but that K, being surrounded by smaller balls such as

4, 5, and 6, only pushes 5 toward ɪ; and at the same time it pushes 4 toward ɪɪ7
L and 6 toward M, and so on. But it does so in such a way that the middle
one, 5, is pushed much more strongly than the others, 4, 6, and similar
ones which are on the sides. In the same way, the ball N, being pushed
toward L, pushes the small balls 1, 2, and 3, one toward L, the other toward
ɪ, and the other toward M, but with the difference that it is 1, and not the
middle one, 2, that is pushed most strongly of all. And what is more, since
the small balls 1, 2, 3, 4, and so on, are all being pushed in this way at the
same time by the other balls N, P, H, P, they prevent one another being
able to go towards L and M as easily as toward the middle, ɪ. And so if the
whole space LIM were full of similar small balls, the rays of their action
would be distributed there in the same way as I have said those of the
comets within the sphere EBG are.

If you object to this that the inequality between the balls N, P, H, P and
1, 2, 3, 4, and so on is much greater than that I have supposed between
the parts of the second element that compose the sphere EBG and those
that are immediately beneath them in the direction of the Sun, I reply
that the only consequence that can be drawn from this is that less refraction must occur in the sphere EBG than in that composed by the balls 1, 2,
3, 4, etc. But since there is in turn some inequality between the parts of
the second element immediately below this sphere EBG and those lower
still in the direction of the Sun, this refraction increases more and more
as the rays penetrate further; so that, when the rays reach the sphere of
the Earth, DAF, the refraction can easily be as great as, or even greater ɪɪ8
than, that of the action by which the small balls 1, 2, 3, 4, and so on are
pushed. For it is very likely that the parts of the second element near this
sphere of the Earth DAF are not any smaller, compared with those near the
sphere EBG, than are those balls 1, 2, 3, 4, etc., compared with the other
balls N, P, H, P.

Appendix 1

The Dioptrics

Discourse 2: Of Refraction

Later on we shall need to know how to determine this refraction quanti-
tatively, and since the comparison I have just used [between the refraction of light and the penetration of a cloth by a tennis ball] enables this to
be understood easily, I believe it is appropriate that I explain it here and
now, and so as to make it easier to understand, I shall speak first about
reflection. Suppose a ball [fig. 19] is struck from A toward B, and at point
B meets the surface of the ground CBE, which prevents it from going
further and causes it to be deflected: let us see in what direction it will go.
But so that we do not get caught up in new difficulties, assume that the
ground is perfectly flat and hard, and that the ball always travels at a
constant speed, both when it descends and rebounds upwards, and let us
94 ignore entirely the question of the power that continues to move it when
it is no longer in contact with the racquet, as well as any effect of its
weight, bulk, or shape. For we are not concerned here to examine it
closely, and none of these things has a bearing on the action of light, which
is what should concern us. We need only note that the power, whatever it
be, which can make the motion of this ball continue, is different from that
which determines it to move in one direction rather than another. This
can be seen readily from the fact that the motion of the ball depends upon
the force with which the racquet has impelled it, and this same force could
have made it move in any other direction just as easily as toward B;
whereas it is the position of the racquet that determines it to tend toward
B, and which could have determined it to tend there in the same way

76

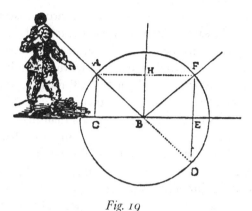

Fig. 19

even if a different force had moved it. This already shows that it is not impossible for this ball to be deflected by its impact with the ground, and hence that there could be a change in its determination to tend toward B without any change in the force of its movement, since these are two different things. And as a result, we must not imagine, as many of our Philosophers do, that it is necessary for the ball to stop for a moment at point B before being reflected toward F; for if its motion were to be interrupted by its being stopped momentarily, there is nothing that would cause it to begin again. Moreover, we should note that not only the determination to move in a particular direction but also the motion itself, and in general any sort of quantity, can be divided into all the parts of which we can imagine it to be composed; and we can readily conceive that the determination of the ball to move from A toward B is composed of two others, one of which makes it descend from line AF to line CE, and at the same time the other makes it go from the left, AC, toward the right, FE, so that the two joined together direct it to B along the line AB. And then it is easy to understand that its impact with the ground can prevent only one of these determinations, and not the other in any way. For it must certainly prevent the one that made the ball descend from AF to CE, for it occupies all the space below CE. But why should it prevent the other, which made the ball move to the right, seeing that it does not offer any opposition at all to the determination in that direction? In order to discover in exactly what direction the ball must rebound, let us describe a circle with centre B passing through point A, and let us say that in as much time as the ball will take to move from A to B, it inevitably returns from B

95

77

to a certain point on the circumference of this circle, inasmuch as all the points which are the same distance away from B as A is can be found on 96 this circumference, and inasmuch as we assume the movement of this ball to be always of a constant speed. Next, in order to determine precisely the point on the circumference to which the ball must return, let us draw three straight lines AC, HB, and FE perpendicular to CE, so that the distance between AC and HB is neither greater nor less than that between HB and FE. And let us say that in the same amount of time as the ball took to move toward the right side from A – one of the points on the line AC – to B – one of those on the line HB – it must also advance from the line HB to some point on FE; for any point of this line FE is as distant from HB in this direction as is any other, as are those on the line AC; and the ball has as much determination to move to that side as it had before. Thus the ball cannot arrive simultaneously both at some point on the line FE and at some point on the circumference of the circle AFD, unless this point is either D or F, as these are the only two points where the circumference and the line intersect; and, accordingly, since the ground prevents the ball from passing toward D, we must of necessity conclude that it inevitably passes toward F. And so it is easy for you to see how reflection occurs, namely at an angle always equal to the one we call the angle of incidence; in the same way that, if a ray coming from point A descends to point B on the surface of a flat mirror CBE, it is reflected toward F in such a way that the angle of reflection FBE is neither greater nor smaller than the angle of incidence ABC.

97 Let us come now to refraction. And first let us suppose that a ball impelled from A toward B encounters at point B, no longer the surface of the earth, but a cloth CBE, which is so thin and finely woven that this ball has the force to rupture it and pass through it, losing only some of its speed: half, for example. Now given this, in order to know what path it must follow, let us consider that its motion is completely different from its determination to move in one direction rather than another, from which it follows that the amounts of these two must be examined separately. And let us also take into account that, of the two parts of which we can imagine this determination to be composed, only the one making the ball tend downwards can be changed in any way as a result of its collision with the cloth, while that making the ball tend to the right must always remain the same as it was, because the cloth offers no resistance to it. Then, having described the circle AFD with its centre at B [fig. 20], and

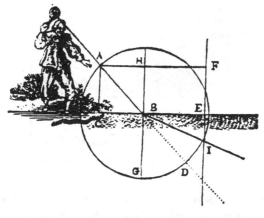

Fig. 20

having drawn at right angles to CBE the three straight lines AC, HB, FE so that the distance between FE and HB is twice that between HB and AC, we shall see that the ball must tend toward point I. For since it loses half its speed in passing through the cloth CBE, it must take twice the time to 98 descend from B to some point on the circumference of the circle AFD as it took to go from A to B above the cloth. And since it loses none of the determination it had to move to the right, in twice as much time as it took to pass from the line AC to HB it must cover twice the distance in the same direction, and consequently it must arrive at a point on the straight line FE at the same instant as it reaches a point on the circumference of the circle AFD. This would be impossible were it not going toward I, as this is the only point below cloth CBE where the circle AFD and the straight line FE intersect.

Now suppose that the ball coming from A toward D does not strike a cloth at point B, but rather a body of water, whose surface slows it by half, as did the cloth. Everything else being as before, I say that this ball must pass from B in a straight line, not toward D, but toward I. For, in the first place, the surface of the water must certainly deflect it there in the same way as did the cloth, seeing that it takes the same amount of force from the ball, and that it offers opposition to the ball in the same direction. Then as for the rest of the body of water which fills the space between B and I, although it may resist the ball more or less than did the air which 99 we previously supposed to be there, this does not mean that it must deflect it more or less. For the water can open up to make way for the ball just as

79

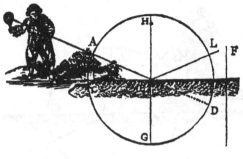

Fig. 21

easily in one direction as in another, at least if we always assume, as we do, that neither the heaviness or lightness of this ball, nor its bulk, nor its shape, nor any other such extraneous cause changes its course. And it may be noted here that the deflection of the ball by the surface of the water or the cloth is greater, depending on how obliquely it enters, so that if it meets it at a right angle, as when it is impelled from H toward B, it must go on in a straight line toward G without any deflection. But if it is impelled along a line such as AB [fig. 21], which is so sharply inclined to the surface of the water or cloth CBE that the line FE, being drawn as before, does not cut the circle AD at all, the ball should not penetrate it at all, but rebound from its surface B toward the air L, in the same way as if it had encountered the earth there. Some have occasionally experienced this with regret when, firing artillery pieces toward the bottom of a river for their amusement, they have wounded those on the other shore.

But let us make another assumption here, and consider that the ball, 100 having first been impelled from A toward B, is impelled again at point B by the racquet CBE, which increases the force of its motion, say by a third, so that afterwards it can progress as much in two moments as it previously did in three. This will have the same effect as if the ball were to meet at point B a body of such a nature that it could pass through its surface CBE one-third again more easily than through the air [fig. 22]. And it follows clearly from what has already been demonstrated that if we describe the circle AD as before, and the lines AC, HB, FE in such a way that there is a third less distance between FE and HB than between HB and AC, then point I, where the straight line FE and the circular one AD intersect, will indicate the place toward which this ball, when it is at point B, must be deflected.

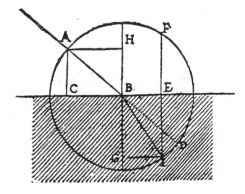

Fig. 22

Now we can also draw the converse of this conclusion and say that since the ball which comes in a straight line from A to B is deflected at B and moves from there toward point I, this means that the force or facility with which it penetrates the body CBEI is to that with which it leaves the body ACBE as the distance between AC and HB is to that between HB and FI, that is, as the line CB is to BE.

Finally, inasmuch as the action of light follows in this respect the same laws as the movement of the ball, it must be stated that when its rays pass obliquely from one transparent body to another which receives them more or less easily than the first, they are deflected in such a way that they are always less inclined to the surface of these bodies on the side of the one that receives them most easily than on the side of the other; and this exactly in proportion to the facility with which the one rather than the other receives them. Only it has to be carefully noted that this inclination must be measured by the size of the straight lines – CB or AH, EB or IG, and the like – in comparison to one another, not by that of angles such as ABH or GBI, and still less by that of angles such as DBI, which we call 'angles of refraction'. For the ratio or proportion between these angles varies with all the different inclinations of the rays; whereas that between the lines AH and IG or other similar ones, remains the same in all refractions caused by the same bodies. Thus, for example [fig. 23], if a ray passing through the air from A to B meets the surface of the lens CBR at point B and is deflected toward I in this lens, and if another ray coming from K toward B is deflected toward L, and another coming from P is deflected toward S, there must be the same ratio between the lines KM and LN, or PQ and ST, as

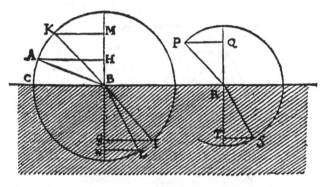

Fig. 23

between AH and IG; but not the same ratio between the angles KBM and LBN, or PRQ and SRT, as between ABH and IBG.

102   So now you see how refractions have to be measured; and although we must refer to observation to determine their quantity, in so far as it depends on the nature of the particular body where they occur, we nevertheless are able to do so with sufficient certainty and facility, since they are all thereby reduced to a common measure. For it is enough to examine a single ray among them to discover all the refractions occurring at a single surface, and we can avoid any error if, in addition, we examine those of several other rays. Thus, if we wish to determine the amount of refraction that occurs at the surface CBR separating the air AKP from the lens LIS, we need only determine the refraction of the ray ABI by examining the proportion between lines AH and IG. Then, if we suspect we have made a mistake in this observation, we must determine the refraction in several other rays like KBL or PRS; and if we find the same proportion between KM and LN, and between PQ and ST, as between AH and IG, we shall have no further cause to doubt the truth of our observation.

    You will, perhaps, be surprised, in carrying out these observations, to find that light rays are more sharply inclined in air than in water, on the surfaces where their refraction occurs, and still more so in water than in glass: which is just the opposite of the ball, which is inclined more sharply 103 in water than in air, and which cannot pass through glass at all. For example, if it is a ball that is impelled through the air from A toward B [fig. 24], meeting a surface of water CBE at point B, it will be deflected from B toward V; yet in the case of a ray, it will, on the contrary, pass from B toward I. However, you shall cease to find this strange when you consider

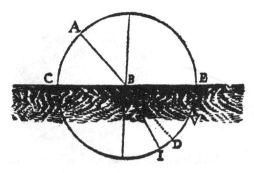

Fig. 24

the nature that I ascribed to light, when I said it is nothing but a certain movement or an action received in the very subtle matter that fills the pores of other bodies; and you should bear in mind that, as a ball loses much more of its agitation in falling against a soft body than against a hard one and rolls less easily on a carpet than on a completely bare table, so the action of this subtle matter can be impeded much more by the parts of the air – which, being as it were soft and badly joined, do not offer it much resistance – than by those of water, which offer it much more; and still more by those of water than by those of glass or crystal. Thus, to the degree that the tiny parts of a transparent body are harder and firmer, the more easily they allow the light to pass; for the light does not have to drive any of them from their places, as a ball must expel those of water if it is to find a passage through them.

Moreover, knowing thus the cause of the refractions that occur in water 104 and glass, and generally in all the other transparent bodies around us, we can note that they must be just the same when the rays are leaving the bodies and when they are entering them. So, if a ray coming from A toward B is deflected from B toward I in passing from the air into a lens, the one that returns from I toward B must also be deflected from B toward A. Nevertheless, other bodies can be found, especially in the sky, where the refractions, proceeding from other causes, are not reciprocal in this way. And certain cases can also be found in which the rays must be curved, even though they pass through only one transparent body, in the same way that the motion of a ball is often curved because it is deflected in one direction by its weight and in another by the action with which we have impelled it, or for various other reasons. For in the final analysis, I dare to say that the three comparisons that I have just used are so fitting

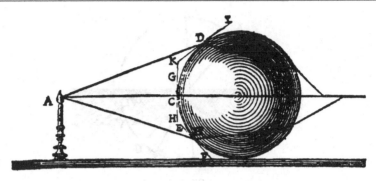

Fig. 25

that all the particular features that may be observed in them correspond to particular features which prove to be entirely similar in the case of light; but I have tried to explain only those that have the greatest bearing on my subject. And I do not wish to have you consider anything else here, except that the surfaces of transparent bodies that are curved deflect the rays passing through each of their points in the same way as would the flat surfaces that we can imagine touching these bodies at the same points. So, for example, the refractions of the rays AB, AC, AD [fig. 25], which come from the flame A and fall on the curved surface of the crystal ball BC, must be regarded in the same way as if AB fell on the flat surface EBF, AC on GHC, and AD on IDK, and likewise for the others. From which you can see that these rays may be variously collected or dispersed, depending on the different curvatures of the surfaces on which they fall. ... [*At this point Descartes introduces the topic of the next Discourse.*]

Appendix 2

The Meteorology

Discourse 8: On the Rainbow

AT VI 325

The rainbow is such a remarkable phenomenon of nature, and its cause has always been so carefully sought after by good minds, yet so little understood, that I could not choose anything better to show you how, by means of the method I am using, we can arrive at knowledge not possessed by any of those whose writings we have. First, taking into consideration that this arc can appear not only in the sky but also in the air near us whenever there are drops of water in it that are illuminated by the Sun – as we can observe in certain fountains – it was easy for me to judge that it came merely from the way in which rays of light act against those drops, and from there tend toward our eyes. Then, knowing that these drops are round, as we have demonstrated above, and seeing that their size does not affect the appearance of the arc, I decided to make a very large [drop] so as to be able to examine it better. For this purpose, I filled a perfectly round and transparent large flask with water, and I discovered that, for example, when the Sun came from the part of the sky marked AFZ [fig. 26], my eye being at point E, when I placed this globe at the spot BCD, its part 326 D appeared to me completely red and incomparably more brilliant than the rest; and I found that whether I approached it or drew back from it, whether I placed it to the right or to the left, or even made it turn around my head, provided that the line DE always made an angle of approximately 42° with the line EM, which one must take to extend from the centre of the eye to the centre of the Sun, this part D always appeared equally red. But as soon as I made this angle DEM slightly larger, the red disappeared.

85

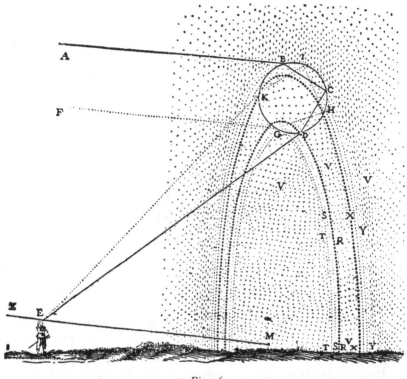

Fig. 26

327 And if I made it slightly smaller, it did not disappear immediately, but
rather divided first into two less brilliant parts, in which yellow, blue, and
other colours were to be seen. Next, looking at the part of the globe
marked K, I perceived that if I made the angle KEM around 52°, this part
K would also appear red, but not as brilliantly as at D; and if I made it
slightly larger, other fainter colours would appear; but I found that if I
made it very slightly smaller still, or very much larger, none at all would
appear. From this, I readily understood that if all the air around M were
filled with such globes – or, in their place, drops of water – a strong red
and very brilliant point must appear in each of those drops from which
lines drawn toward the eye E make an angle of around 42° with EM, as I
assume to be the case with those marked R. And if these points are viewed
all together, without our noting anything about their position except the
angle at which they are seen, they must appear as a continuous band of
red. And in the same way there must be points in those drops marked S

86

and T from which lines drawn to E make slightly more acute angles with EM, and these points make up bands of weaker colours. And in this consists the primary or main rainbow. And then again I found that if the angle MEX was 52°, a red band must appear in the drops marked X, and other bands of fainter colours in the bands marked Y, and that this is what the secondary or minor rainbow consists in. And finally, no colours at all 328 can appear in all the other drops, marked V. After I had done this, I examined in more detail what caused the part D of the globe BCD to appear red, finding that it was the rays of the Sun which, coming from A toward B, were bent on entering the water at point B, and went toward C, from where they were reflected toward D; and there, being bent again as they left the water, they tended toward E. For when I put an opaque or dark body in some place on the lines AB, BC, CD, or DE, this red colour would disappear. And even if I covered the whole globe, except for the two 329 points, and put dark bodies everywhere else, provided that nothing hindered the action of the rays ABCDE, the red colour appeared nevertheless. Then, looking also for the cause of the red that appeared at K, I found that it was the rays coming from F toward G, which were bent there toward H, and at H they were reflected toward I, at I it was reflected again toward K, and then finally bent at point K and tended toward E. So that the primary rainbow is caused by the rays reaching the eye after two refractions and one reflection, and the secondary by rays reaching it only after two refractions and two reflections; which is what prevents the second appearing as clearly as the first.

But the principal difficulty still remained: namely, to understand why, when there are many other rays there which, after two refractions and one or two reflections, can tend toward the eye when the globe is in a different position, it is nevertheless only those of which I have spoken that cause certain colours to appear. To resolve this difficulty, I looked to see if there was something else in which they appeared in the same way, so that by comparing these with each other I would be in a better position to gauge their cause. Then, remembering that a prism or triangle of crystal causes similar colours to be seen, I considered one of them such as MNP [fig. 27], which has two completely flat surfaces, MN and NP, inclined to one another at an angle of around 30° or 40°, so that if the rays of the Sun ABC cross MN at right angles, or almost at right angles, so that they do not 330 undergo any noticeable refraction there, they must suffer a reasonably large refraction on leaving through NP. And when I covered one of these

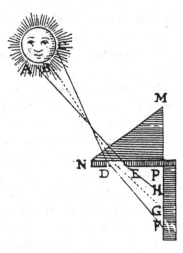

Fig. 27

two surfaces with a dark body, in which there was a rather narrow opening DE, I observed that the rays, passing through this opening and from
there making for the cloth or paper FGH, paint all of the colours of the
rainbow on it; and that they always paint the colour red at F, and blue or
violet at H. From this I learned, first, that the surfaces of the drops of
water do not need to be curved in order to produce these colours, as those
of this crystal are completely flat. Nor does the angle under which they
appear need to be of any particular size, for this can be changed without
any change in them, and although we can make the rays travelling toward
F bend more or less than those travelling toward H, they nevertheless
always colour it red, and those going toward H always colour it blue. Nor
is reflection necessary, for there is no reflection here, nor finally do we
need many refractions, for there is only one refraction here. But I
reasoned that there must be at least one refraction – and, in fact, one
whose effect was not destroyed by another – for experience shows that, if
the surfaces MN and NP are parallel, the rays, being straightened as much
331   in one as they were able to be bent in the other, would not produce these
colours. I did not doubt that light was also needed, for without it we see
nothing. And moreover, I observed that shadow, or some limitation on
this light, was necessary; for if we remove the dark body from NP, the
colours FGH cease to appear; and if the opening DE is made large enough,
the red, orange, and yellow at F reach no further because of that, any more

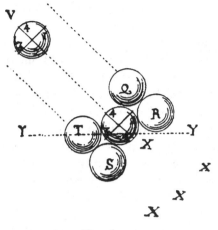

Fig. 28

than do the green, blue, and violet at H. Instead, all the extra space at G between these two remains white. After this, I tried to understand why these colours are different at H and at F, even though the refraction, shadow, and light combine there in the same way. And conceiving the nature of light to be such as I described it in the *Dioptrics*, namely as the action of motion of a certain very subtle matter, whose parts should be imagined as small balls rolling in the pores of terrestrious bodies, I understood that these balls can roll in different ways, depending on the causes that determine them; and in particular that all the refractions that occur on the same side cause them to turn in the same direction. But when they have no neighbouring balls that move significantly faster or slower than they, their rotation is approximately equal to their rectilinear motion, whereas when they have some on one side that move more slowly, and others on the other side that move as fast or faster, as happens when they are bounded by shadow and light, then when they encounter those which are moving more slowly on the side toward which they are rolling, as do those making up the ray EH, this causes them to rotate less quickly than if they were moving in a straight line. And the opposite happens when they encounter them on the other side, as do those of ray DF. To understand this better, imagine the ball 1234 [fig. 28] being propelled from v toward x, in such a way that it travels only in a straight line, and that its two sides 1 and 3 descend equally quickly toward the surface of the water YY, where the movement of the side marked 3, which encounters it first, is retarded, 332

89

while that of side ı still continues; this causes the whole ball to begin inexorably to rotate following the numbers 1234. Then, imagine it is surrounded by four others – Q, R, S, T – of which Q and T tend to move toward x with a greater force than does the ball, and the other two – s and T – tend there with less force. It is clear from this that Q, which presses
333 the part of the ball marked ı, and s, which retains that marked 3, increase its rotation; and that R and T do not hinder it, because R is disposed to move toward x faster than the ball follows it, and T is not disposed to follow the ball as quickly as it precedes it. This explains the action of the ray DF. And on the other hand, if Q and R tend more slowly than it toward x, and s and T tend there more rapidly, R hinders the motion of that part marked ı, and T that of part 3, without the two others – Q and s – doing anything. This explains the action of the ray EH. But it is worth noting that since this ball 1234 is quite round, it can easily happen that, when it is pressed hard by the two balls R and T, it is turned and rotates around the axis 42, rather than their causing its rotation to stop. And so, changing its position in an instant, it subsequently rotates following the numbers 321; for the two balls R and T, which caused it to begin to rotate, make it continue until it has completed a half-turn in this direction, and then they can increase its rotation instead of retarding it. This enabled me to resolve the major difficulty that I had in this matter. And it seems to me that it is very evident from all of this that the nature of the colours appearing at F consists just in the parts of the subtle matter which transmit the action of light having a much greater tendency to rotate than to travel in a straight line. As a consequence, those which have a much stronger tendency to rotate cause the colour red, and those which have only a slightly stronger
334 tendency cause yellow. The nature of those that are visible at H, on the other hand, consists just in the fact that these small parts do not rotate as quickly as normal, when there is no particular cause hindering them; so that green appears when they rotate just a little more slowly, and blue when they rotate very much more slowly. And usually this blue is combined with a pinkish colour at its edges, which makes it vivacious and sparkling, and changes it into violet or purple. The cause of this is without doubt the same as that which usually slows down the rotation of the parts of the subtle matter when it has enough strength to change the position of some of them and increase their rotation, while slowing that of others. And the explanation agrees so well with observation in all of this that I do not believe it possible, after one has attended carefully to

both, to doubt that things are such as I have explained. For if it is true that
the sensation we have of light is caused by the movement or inclination
to movement of some matter touching our eyes, as is indicated by many
other things, it is certain that different movements of this matter must
cause different sensations in us. And as these movements cannot differ
other than in the way I have mentioned, we observe no difference in the
sensations we have of them other than a difference in colour. And we can
find nothing at all in the crystal MNP that can produce colours except the
way in which it sends the tiny bits of subtle matter toward the line FGH, 335
and from there toward our eyes. From this, it seems to me obvious enough
that we should not look for anything else in the colours that other objects
make appear; for ordinary observation shows that light or white, and
shadow or black, together with the colours of the rainbow that have been
explained here, are enough to make up all the others. And I cannot accept
the distinction the Philosophers make between true colours and others
which are only false or apparent. For because the entire true nature of
colours consists only in their appearance, it seems to me to be a contradiction to say both that they are false and that they appear. But I acknowledge that shadow and refraction are not always necessary to produce
them, and that instead of these, the size, shape, situation, and movement
of the parts of the bodies that one terms 'coloured' can combine in
various ways with light to increase or diminish the rotation of the parts of
the subtle matter. So, even in the rainbow, I initially doubted whether the
colours there were produced in the same way as in the crystal MNP; for I
did not notice any shadow cutting off the light, nor did I yet understand
why they appeared at different angles, until, having taken my pen and
calculated in detail all the rays that fall on the various points on a drop of 336
water, in order the find out at what angles they would come to our eyes
after two refractions and one or two reflections, I found that after one
reflection and two refractions, far more of them can be seen at the angle
of 41° to 42° than at any smaller one; and that none of them can be seen
at a larger angle. Next I also found that after two reflections and two
refractions, far more of them come to the eye at an angle of 51° to 52° than
at any greater one; and no such rays come at a smaller one. So that there
is a shadow on both sides, cutting off the light which, having passed
through innumerable raindrops illuminated by the Sun, comes toward
the eye at an angle of 42° or slightly less, thus causing the primary, main
rainbow. And there is also one cutting off the light at an angle of 52° or

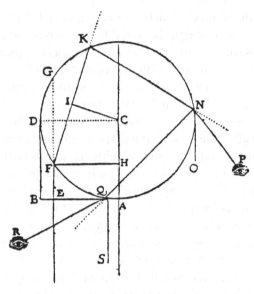

Fig. 29

slightly more, which causes the outer rainbow; for failing to receive rays of light in your eyes, or receiving very much fewer of them from one object than from another which is near it, is to see a shadow. This is a clear demonstration that the colours of these arcs are produced by the same cause as those that appear with the aid of the crystal MNP, and that the radius of the inner arc must not be greater than 42°, nor that of the outer one less than 51°, and finally that the outside surface of the primary rainbow must be much more restricted than the inside one; and the opposite in the case of the secondary one, as observation shows us. But so that those who have a knowledge of mathematics can understand whether the calculation I have made of these rays is sufficiently exact, I should explain it here.

337

Let AFD be a drop of water [fig. 29] whose radius CD or AB I divide into as many equal parts as I wish to calculate rays, so as to attribute an equal amount of light to them all. Then I consider one of these rays in detail: EF, for example, instead of passing directly through G, is deflected toward K, is reflected from K toward N, from where it goes toward the eye P; or alternatively it is reflected once more from N to Q, and from there is turned toward the eye R. And having drawn CI at right angles on FK, I know from what was said in the *Dioptrics* that the ratio between AE (or HF)

92

and CI is that by which the refraction of water is measured. So that if HF contains 8,000 parts – taking AB to contain 10,000 – CI will contain around 5,984 because the refraction of water is slightly greater than 3/4. On the most exact measurement I have been able to make, it is 187/250. Having thus the two lines HF and CI, I could easily find the size of the two arcs, FG, which was 73° 44′, and FK, which was 106° 30′. Then, taking double the arc FK from the arc FG added to 180°, I obtain 40° 44′ for the size of the angle ONP, on the assumption that ON is parallel to EF. And taking this 40° 44′ from FK, I have 65° 46′ for the angle SQR, assuming that SQ is parallel to EF. And doing the same calculation for all the other rays parallel to EF which pass through the divisions of the diameter AB in the same way, I compile table [1]. 338

It can be readily seen from table [1] that there are many more rays making the angle ONP around 40° than there are those making it less; and also more of them that make SQR around 54° than make it larger. Then, so as to make it still more precise, we have [see table 2]. 339

And I see here that the largest angle, ONP, can be 41° 30′, and the smallest, SQR, 51° 54′; when I add or subtract around 17′ for the radius of the Sun, I have 41° 47′ for the largest radius of the inner rainbow, and 51° 37′ for the smallest radius of the outer one. 340

It is true that when the water is warm its refraction is slightly less than when it is cold, which can alter certain things in the calculation. Nevertheless, it will only increase the radius of the inner rainbow by one or two degrees at the most, in which case that of the outer rainbow will be nearly twice that of the smaller. This is worth noting because by these means we can demonstrate that the refraction of water can be hardly any more or less than I have supposed. For if it were slightly larger, it would make the radius of the inner rainbow less than 41°, whereas the common belief is that it is 45°; and if we assume it to be small enough to make it exactly 45°, we will find that the radius of the outer arc is also hardly more than 45°, whereas it appears much larger to the eye than the inner one. And Macrolius, who I believe is the first to have determined the figure of 45°, determines the other to be around 56°, which shows how little faith we should have in observations which are not accompanied by true rationale. For the rest, I have had no difficulty in understanding why red is on the outside of the inner arc, nor why it is on the inside on the outer one; for the same thing that causes it to be near F rather than H when it appears through the crystal MNP, also causes us to see the red toward its 341

The World and Other Writings

Table 1

Line HF	Line CI	Arc FG	Arc FK	Angle ONP	Angle SQR
1000	748	168°30′	171°25′	5°40′	165°45′
2000	1496	156°55′	162°48′	11°19′	151°29′
3000	2244	145°4′	154°4′	17°56′	136°8′
4000	2992	132°5′	145°10′	22°30′	122°4′
5000	3740	120°	136°4′	27°52′	108°12′
6000	4488	106°16′	126°40′	32°56′	93°44′
7000	5236	91°8′	116°51′	37°26′	79°25′
8000	5984	73°44′	106°30′	40°44′	65°46′
9000	6732	51°41′	95°22′	40°57′	54°25′
10000	7480	0°	83°10′	13°40′	69°30′

thicker part MP, and blue toward N, when we look at this crystal with our eye at the white screen FGH, because the ray coloured red which goes toward F comes from C, the part of the Sun that is closest to MP. And this same cause brings it about that, when the centre of the drops of water – and as a result their broadest part – are on the outside of the coloured points forming the interior rainbow, the red must appear there on the outside; and that when they are on the inside of those that make up the outer rainbow, the red must correspondingly appear on the inside.

Thus I believe that there remains no difficulty in this matter, unless it perhaps concerns the irregularities which one encounters here: for example, when the arc is not exactly round, or when its centre is not in a straight line passing through the eye and the Sun, which can occur if the wind changes the shape of the raindrops. For losing the smallest part of their roundness must make a significant difference in the angle at which the colours must appear. I have been told that there has sometimes also been observed a rainbow so reversed that its ends were turned upwards, as represented here [fig. 30] at FF. I can only account for this in terms of the reflection of the rays of the Sun falling on the water of the sea or some lake. Assume they come from the part of the sky SS, fall on the water DAE, and from there are reflected toward the rain CF; then the eye B will see the arc FF, whose centre is at point C, so that if CB is projected right to A, and

342

94

Table 2

Line HF	Line CI	Arc FG	Arc FK	Angle ONP	Angle SQR
8000	5984	73°44′	106°30′	40°44′	65°46′
8100	6058	71°48′	105°25′	40°58′	64°37′
8200	6133	69°50′	104°20′	41°10′	63°10′
8300	6208	67°48′	103°14′	41°20′	62°54′
8400	6283	65°44′	102°9′	41°26′	61°43′
8500	6358	63°34′	101.2′	41°30′	60°32′
8600	6432	61°22′	99°56′	41°30′	58°26′
8700	6507	59°4′	98°48′	41°28′	57°20′
8800	6582	56°42′	97°40′	41°22′	56°18′
8900	6657	54°16′	96°32′	41°12′	55°22′
9000	6732	51°41′	95°22′	40°45′	54°20′
9100	6806	49°0′	94°12′	40°36′	53°36′
9200	6881	46°8′	93°2′	40°4′	52°58′
9300	6956	43°8′	91°51′	39°26′	52°52′
9400	7031	39°45′	90°38′	38°38′	52°0′
9500	7106	36°24′	89°26′	37°32′	51°54′
9600	7180	32°30′	88°12′	36°6	52°6′
9700	7255	28°8′	86°58′	34°12	52°46′
9800	7330	22°57′	85°43′	31°31′	54°12′

AS passes through the centre of the Sun, the angles SAD and BAE are equal, and so the angle CBF is around 42°. Nevertheless, for this effect to occur there must also be absolutely no wind to disturb the surface of the water at E, and there must also perhaps be a cloud such as G which prevents the Sun's light, which travels in a straight line toward the rain, from effacing the light that this water E sends there. Consequently, this is a rare occurrence. Besides, it is possible for the eye to be in such a position with respect to the Sun and the rain that the lower part, where the band of the rainbow is terminated, is seen, but not the upper part; and then it will be taken to be an inverted arc, even though we do not see it near the sky, but near the water or the earth.

I have also been told that a third rainbow has sometimes been seen 343

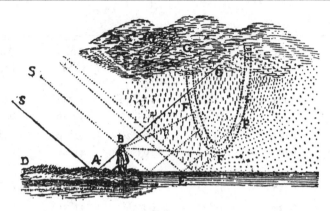

Fig. 30

above the two usual ones, but that it was much fainter, and approximately as distant from the secondary one as that is from the primary. I do not think this could have happened unless there had been numerous round and transparent grains of hail mixed in with the rain. Since the refraction in these is significantly greater than that in water, the outer rainbow must have been very much larger there, and so appears above the others. As for the inner one, which for the same reason would have to have been smaller than that of the rain, it possibly will not have been noticed, because of the great lustre of the outer one. Or alternatively, because their edges are joined, the two of them will be counted as one, but one whose colours are arranged differently than is usual.

[*The final paragraph of Discourse 8 describes how to produce optical illusions with fountains.*]

The *Treatise on Man* and related material

Treatise on Man

[Part 1: On the machine of the body][1]

These men[2] will be composed, as we are, of a soul and a body. And I must describe for you first the body on its own; and then the soul, again on its own; and finally I must show you how these two natures would have to be joined and united so as to constitute men resembling us.

I suppose the body to be just a statue or a machine made of earth,[3] which God forms with the explicit intention of making it as much as possible like us. Thus He not only gives its exterior the colours and shapes of all the parts of our body, but also places inside it all the parts needed to make it walk, eat, breathe, and imitate all those functions we have which can be imagined to proceed from matter and to depend solely on the disposition of our organs.

We see clocks, artificial fountains, mills, and other similar machines which, even though they are only made by men, have the power to move of their own accord in various ways. And, as I am supposing that this machine is made by God, I think you will agree that it is capable of a greater variety of movements than I could possibly imagine in it, and that it exhibits a greater ingenuity than I could possibly ascribe to it.

I shall not pause to describe to you the bones, nerves, muscles, veins, arteries, stomach, liver, spleen, heart, brain, nor all the other different parts from which this machine must be composed, for I am assuming

---

[1] The division into four parts and the provision of titles are due to Clerselier, not Descartes. Clerselier also divided the text into 106 articles: see AT x. 203–9.

[2] This presumably refers to some men mentioned in some projected or missing section which came either at the beginning of 'Man' or at the end of the *Treatise on Light*.

[3] That is, the element earth, as described in ch. 5 of the *Treatise on Light*, above pp. 16–21.

that they are just like those parts of our own bodies having the same
names, and that you can get some learned anatomist to show them to
you – at least those which are large enough to be seen with the naked
eye – if you are not already sufficiently acquainted with them. And as for
those that are too small to be seen, I can let you know about them most
readily by telling you of the movements that depend on them; so that
it remains only for me to explain these movements to you here in the
proper order and by these means to tell you which of our functions these
represent.

First, food is digested in the stomach of this machine by the force of
certain fluids which, gliding among its parts, separate, shake, and heat
them, just as ordinary water does those of quicklime, or *aqua fortis* those
of metals. Furthermore these fluids, since they are brought from the heart
through the arteries very quickly, must be very hot, as I shall explain
below. And the food is usually of such a nature that it can be broken down
and heated up of itself, just as occurs with new hay if it is shut up in the
barn before it is dry.[4]

It should also be pointed out that the agitation which is induced in the
small particles of food when they are heated, together with the agitation
of the stomach and the bowels in which they are contained, as well as the
arrangement of the fibres from which the bowels are composed,[5] cause
these particles, to the extent to which they are digested, to descend
gradually toward the passage through which the coarsest of them must
exit.[6] And the finest and most agitated meanwhile encounter innumerable
small holes through which they flow into the branches of a large vein that
bears them toward the liver, and into others that bear them elsewhere,
with nothing but the small size of the holes serving to separate these from
the coarser particles; just as, when one shakes meal in a sieve, the purest
parts flow out and it is only the small size of the holes through which it
passes that prevents the bran from following after them.

These finer parts of the food, being of different sizes and still imper-
fectly mixed together, make up a fluid which would remain quite agitated
and whitish were it not that a part of it is blended straightaway with the

---

[4] The comparison with thermogenic processes is crucial to Descartes' account, and he considers
such processes in the *Discourse on Method* (Part 5), the *Meteors* (Discourse 7), and in the *Principles*
(Part 4, art. 92).
[5] Descartes assumes (following a long medical tradition) that fibres are the fundamental structural
elements in the body generally.
[6] Descartes is referring to the rectum here.

mass of blood that is contained in the branches of what is called the portal vein (which receives this fluid from the intestines), in what is called the vena cava (which conducts it toward the heart), and in the liver itself 123 as if it were a single vessel.[7]

Similarly, it should be noted here that the pores of the liver are arranged in such a way that this fluid, on entering, is refined and transformed, taking on the colour and form of blood, just as the white juice of black grapes is converted into light-red wine when it is allowed to ferment on the vine stock.[8]

Now there is only one passage evident by which this blood, thus contained in the veins, can leave them, namely that which conveys it to the right cavity of the heart. And note that the flesh of the heart contains in its pores one of those fires without light which I have spoken about earlier and which makes it so fiery and hot[9] that, to the extent that the blood enters either of its two chambers or cavities, it is promptly inflated and expanded. Similarly, it can be demonstrated experimentally that the blood or milk of some animal will be dilated if you pour it a drop at a time into a very hot flask. And the fire in the heart of this machine that I am describing to you has as its sole purpose to expand, warm, and refine the blood that falls continually a drop at a time through the passage from the vena cava into the cavity on its right side, from where it is exhaled into the lung, and from the vein of the lung which anatomists have called the 'venous artery'[10] into its other cavity, from where it is distributed throughout the body.

The flesh of the lung is so rare, so soft, and always so refreshed by the air from respiration, that as the blood vapours, which go out from the right cavity of the heart, enter it through the artery that anatomists call 124

---

[7] In an important letter to Regius of 24 May 1640, which is a commentary on Regius's account of Cartesian physiology, Descartes says that the food first becomes *chyle* in the stomach, then *chyme* in the liver, as the result of a kind of fermentation, and finally *blood*, in the heart, as a result of an ebullient reaction (AT iii. 66–8). Surprisingly, this account does not figure in the *Description of the Human Body* (see below), which was composed after 1640.

[8] The comparison between the processes of sanguinification and fermentation was traditional, and can be found in Galen (*De usu partium*, bk. 4, ch. 3), but Descartes' account of the nature of the process involved, which offers a corpuscularian reduction, is very different from the traditional account.

[9] The idea that the heart heats and transforms the blood goes back to Aristotle and Galen, and was very widely, if not universally, accepted before Harvey, as was the idea that the blood acts as a kind of fuel for the heart. Harvey argued that it is the blood that heats the heart, not vice versa.

[10] Descartes will point out in the *Description of the Human Body*, correctly, that this is really a vein – the pulmonary vein – although he will continue to use the standard anatomical term 'venous artery'.

the 'arterial vein'[11] they are thickened and converted back into blood again. This blood then falls a drop at a time into the left cavity of the heart where, if it were to enter it without being thickened again, it would be inadequate to sustain the fire that is there.

Thus you see that respiration, whose sole purpose in this machine is to thicken the vapours, is as necessary for maintaining the fire in its heart as it is in us for maintenance of our life, at least in those of us who are fully formed: for in infants who are still in their mothers' wombs, and so unable to draw in fresh air by respiration, two passages make up for this. Blood passes through one of these from the vena cava to the vein which is called an artery [the pulmonary vein], while through the other the vapours or rarefied blood are breathed out from the artery which is called a vein [the pulmonary artery] and enter the great artery [the aorta]. And in the case of animals that have no lung at all, they have only one cavity in their heart or, where they have several, they are all in a single sequence.

The pulse, or the beating of the arteries, depends on eleven small membranes[12] which, like so many small doors, close and open the orifices of the four vessels that open into the two cavities of the heart. For at the moment when a beat ceases and another one is ready to begin, the small doors at the orifices of the arteries are shut tight, while those at the orifices of the two veins are open, so that two drops of blood cannot but fall immediately from these two veins, one into each cavity of the heart. These drops of blood, being rarefied and suddenly occupying a space which is incomparably greater than that which they occupied previously, then push the small doors at the orifices of the veins shut, thereby preventing more blood from dropping into the heart, and they push open the arteries, passing through them quickly and forcefully, and cause the heart and all the body's arteries to inflate at the same time. But immediately after this, this rarefied blood is either condensed again or penetrates other parts; and thus the heart and the arteries are deflated, the small doors at the entrances to the two arteries are shut again, and those at the entrances to the two veins are reopened to allow in two more drops of blood, which cause the heart and the arteries to be inflated again, just as before.

Once we know the cause of the pulse, we can readily understand that

---

[11] See the above note. What the anatomists called the 'arterial vein' is in fact an artery – the pulmonary artery – which is the term I shall use from now on.
[12] These are the cardiac valvules, which had been known to anatomists since Galen. They are described in *Description of the Human Body*, AT xi. 228–30, pp. 172–4 below.

it is not so much the blood contained in the veins of this machine, which has newly come from its liver, but rather the blood contained in the arteries, which has already been distilled in its heart, that is able to attach itself to other parts and can be used to replace what the continual agitation of these parts, not to mention the various actions of surrounding ones, detaches and extricates from them. For the blood in the veins always flows gradually from their extremities toward the heart (and the arrangement of certain little doors or valves which the anatomists have noticed in several places along the veins is enough to convince you that the same 126 happens in us). The blood in the arteries, on the other hand, is pushed out of the heart under pressure and in separate little spurts, towards their extremities. Thus this blood can easily come to join and unite with all the bodily parts, being able to maintain them and even make them grow if the machine represents a person's body which is disposed in the right way.

For at the moment when the arteries inflate, the small parts of blood which they contain will randomly strike the roots of certain little threads which, originating from the extremities of the little branches of these arteries, make up bones, flesh, membranes, nerves, the brain, and all the other solid parts depending on the different ways in which they are joined or interconnected. And thus they have the force to push them in front of them a little, and in this way gradually replace them. Then, at the moment when the arteries deflate, each of these parts is stopped in its place, and this alone means it is joined to those it touches, in accord with what I said above.

Now if our machine represents the body of an infant, its matter will be so tender and its pores so easily enlarged that the parts of the blood that enter in this way into the composition of its solid parts will usually be a little larger than those which they replace; it can even come about that two or three will replace a single one, and this will cause growth. But the matter of its parts will gradually harden in the meantime, so that after a few years its pores will no longer be able to enlarge to the same degree; and 127 so, ceasing to grow, they will represent the body of an older person.[13]

Moreover, only a very few of the parts of the blood can be united on every occasion to the solid parts in the way I have just explained; most of them return through the veins from the extremities of the arteries, which in many places are joined to the extremities of the veins. And perhaps

[13] Compare *Description of the Human Body*, AT xi. 250, p. 185 below.

some parts also pass out of the veins to nourish some of the bodily parts, but the majority return into the heart, and from there go to the arteries again, in such a way that the movement of the blood in the body is just a perpetual circulation.

In addition, there are some parts of the blood that proceed into the spleen, and others to the gall bladder, and, via the spleen and the gall bladder as well as directly from the arteries, there are some parts that re-enter the stomach and the bowels, where they act like *aqua fortis*, helping in the digestion of food. And because they are carried here from the heart almost instantaneously through the arteries, they are always very hot, which enables their vapours to rise easily through the gullet toward the mouth, where they make up the saliva. There are also some that flow out as urine through the flesh of the kidneys, and as sweat or other excrements through the skin. And through whichever of these places it passes, either the position, shape, or smallness of the pores through which they pass is what alone makes some go through and not others, and keeps the rest of the blood from following, just as you see in various sieves which, being pierced in different ways, serve to separate different grains from one another.

128

But what must be noted above all at this point is that all the most energetic, strongest, and finest parts of this blood proceed to the cavities of the brain, inasmuch as the arteries bearing them there are in the most direct line from the heart; and as you know, all moving bodies tend as much as they are able to continue their motion in a straight line.

Consider the heart A, for example [fig. 31], and consider that when the blood is forced from it through the aperture B, all its parts tend toward C, that is, toward the cavities of the brain; but because the passage is not sufficiently large to bear all of them there, the weakest are turned back by the strongest, which in this way proceed there alone.

You should also note in passing that the strongest and most energetic parts, other than those which go directly to the brain, go to the vessels destined for reproduction. For if those that have the force to reach D, for example, cannot progress on to C, because there is no room for all of them there, they turn instead toward E, rather than toward F or G, in so far as the passage toward E is straighter. Beyond this, I could perhaps show you how, from the humour that gathers at E, another machine which is similar to this can be formed, but I do not wish to enter further into this matter.

129

As for those parts of the blood that penetrate as far as the brain, they

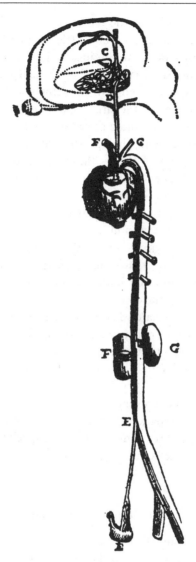

Fig. 31

serve not only to nourish and sustain its substance, but above all to produce there a certain very fine wind, or rather a very lively and very pure flame, which is called the 'animal spirits'. For it should be noted that the arteries that carry these from the heart, after having divided into countless small branches and having composed the little tissues that are stretched out like tapestries at the bottom of the cavities of the brain, come together again around a certain little gland which lies near the middle of the substance of the brain, just at the entrance to its cavities; and those in this region have a large number of small holes through which the finest parts of the blood can flow into this gland, and these are so narrow that they do not allow the larger ones to get past.

You should also know that these arteries do not stop there, but being gathered up into a single one, they go straight up and enter that great vessel which, like Euripos,[14] bathes the whole external surface of the brain.[15] And it must also be noted that the coarsest parts of the blood can lose a lot of their agitation in the twists and turns of the little tissues through which they pass, to the extent that they have the power to push the smaller ones among them and so transfer some of their motion to them; but these smaller ones cannot lose their motion in this way, because the agitation is increased by that which the larger ones transfer to them, and because there are no other bodies around them to which they can transfer theirs with the same ease.

130

It can be readily appreciated from this, that when the coarsest parts go up straight to the external surface of the brain, where they serve to provide nourishment for its substance, they make the smallest and most agitated parts move out of the way, causing all of them to enter this gland, which we must imagine as a very full-flowing spring, and from this they flow at the same time and in every direction into the cavities of the brain. And so, without any preparation or alteration, except being separated from the larger parts and retaining the extreme speed that the heat of the heart has given them, they cease to have the form of blood and are called animal spirits.

[Part 2: How the machine of the body is moved]

Now as these spirits enter the cavities of the brain, they also pass in the same proportions from there into the pores of its substance, and from these pores into the nerves. And depending on which of these nerves they enter, or even merely tend to enter, in varying amounts, they have the

---

[14] The straits of Euripos, which separate the island of Euboea from the Greek mainland, had notoriously reversible tidal currents.

[15] Compare this with the rather different account given in the *Description of the Human Body* (AT xi. 269–70), pp. 195–6 below.

power to change the shapes of the muscles into which these nerves are embedded, and in this way to move all the limbs. Similarly, you may have observed in the grottoes and fountains in the royal gardens[16] that the force that drives the water from its source is all that is needed to move various machines, and even to make them play certain instruments or pronounce certain words, depending on the particular arrangements of the pipes through which the water is conducted.

And the nerves of the machine that I am describing can indeed be compared to the pipes in the mechanical parts of these fountains, its muscles 131 and tendons to various other engines and springs which serve to work these mechanical parts, its animal spirits to the water that drives them, the heart with the source of the water, and the brain's cavities with the apertures.[17] Moreover, respiration and similar actions which are normal and natural to this machine, and which depend on the flow of spirits, are like the movements of a clock or mill, which the normal flow of water can make continuous. External objects, which by their mere presence act on the organs of sense and thereby cause them to move in many different ways,[18] depending on the arrangement of the parts of the brain, are like strangers who on entering the grottoes of these fountains unwittingly cause the movements that take place before their eyes. For they cannot enter without stepping on certain tiles which are arranged in such a way that, for example, if they approach a Diana bathing they will cause her to hide in the reeds, and if they move forward to pursue her they will cause a Neptune to advance and threaten them with his trident; or if they go in another direction they will cause a sea monster to emerge and spew water in their faces; or other such things depending on the whim of the engineers who constructed them. And finally, when a rational soul is present in this machine it will have its principal seat in the brain and will reside there like the fountaineer, who must be stationed at the tanks to which the fountain's pipes return if he wants to initiate, impede, or in 132 some way alter their movements.[19]

---

[16] Descartes is almost certainly referring to the Royal Gardens at Saint-Germain-en-Lay, just outside Paris, with fountains designed by the Fancini brothers. The gardens are illustrated and described by Salamon de Caus in *Des Raisons des forces mouvantes* (1615) and in other contemporary writers.

[17] The *regards*, which I have translated as 'apertures', are inspection holes made in the machine containing the flowing water.

[18] Descartes had begun to describe this process as early as the *Rules*: see especially Rule 12.

[19] This image of the rational mind comes dangerously close to the idea of the mind as being like a pilot guiding a ship, which Descartes will reject in very firm terms in the *Meditations*.

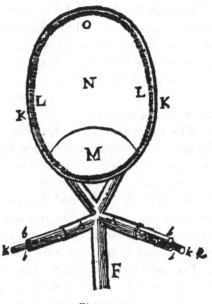

Fig. 32

But so that you might have a firm grasp of all this, I want first to tell you about the composition of the nerves and the muscles, and to show you how, from the sole fact that the spirits in the brain are ready to enter into certain of the nerves, they have the power to move certain bodily parts at the same instant. Then, after touching briefly on respiration and other similar, simple, and normal movements, I shall say how external objects act upon the sense organs. And after this, I shall explain in detail all that happens in the cavities and pores of the brain, what route the animal spirits follow there, and which of our functions this machine can imitate by these means. For if I were to begin with the brain and simply follow in order the route of the spirits, as I did with the blood, it seems to me that what I have to say would be far less clear.

133   Observe, for example, nerve A [fig. 32], whose external membrane is like a large tube which contains several other tiny tubes *b*, *c*, *k*, *l*, etc. made up from a finer internal membrane; and observe also that these two membranes are continuous with the two, K and L, that cover the brain MNO.

Note also that in each of the tiny tubes there is a kind of marrow made up from several very fine fibres which come from the brain's own substance, N, and whose extremities end on the one side at its internal surface facing its cavities, and at the other side at the membranes and the flesh on which the tube containing them terminates. But because this marrow is not used to move the bodily parts, it is enough for the present to note that it does not completely fill the tubes that contain it, but leaves sufficient room for the animal spirits to flow easily through them from the brain into the muscles, to where these tubes, which must be thought of as so many little nerves, proceed.

Next observe [fig. 33 and figs. 34a and 34b][20] how the tube or tiny nerve *bf* proceeds to muscle D, which I assume to be one of those that move the eye, and how it divides there into several branches, composed of a relaxed 134 membrane which can be extended, enlarged, and shrunk depending on the amount of animal spirits that enter or leave it; and its branches or fibres are arranged in such a way that when animal spirits enter there they cause[21] the whole body of the muscle to inflate and shorten and so pull the eye to which it is attached, whereas when they withdraw, on the other hand, the muscle deflates and elongates again.

Observe, moreover [fig. 33], that as well as the tube *bf* there are yet others, namely *ef*, through which the animal spirits can enter muscle D, and *dg*, through which they can leave it. And in just the same way muscle E, which I assume is used to move the eye in the opposite direction, receives animal spirits from the brain through tube *cg* and from muscle D 135 through *dg*, moving them back toward D through *ef*. And note that although there is no passage evident through which the spirits contained in muscles D and E can leave them, except to enter each other, nevertheless because their parts are very small and indeed are made constantly smaller by the force of their agitation, some always escape through the membranes and flesh of the muscles, while others return through the two tubes *bf* and *cg* in a compensatory motion.

Finally, note that, where the two tubes *bf* and *ef* join, there is a certain small membrane H*fi* [fig. 35] that separates these two tubes and acts as a door. It has two flaps, H and *i*, which are arranged in such a way that when the animal spirits tending downwards from nerve *b* towards flap H are stronger than those tending upwards from the muscle E toward flap *i*, they push down on and open this membrane, thus allowing the animal spirits

---

[20] These are alternative figures. Figs. 34a and 34b are based on Descartes' own drawing: see AT xi. 134 note a.

[21] Reading 'font' for 'sont' (AT), which is a misprint.

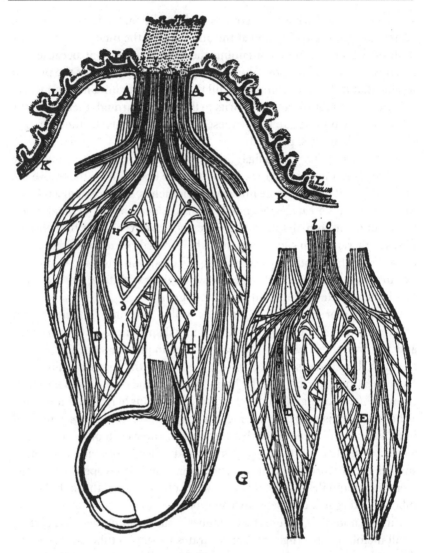

Fig. 33

in muscle E to flow promptly toward D. And when those that tend from *e* toward *i* are stronger, or even when they are just of the same strength as the others, they raise and close H*fi* and thus prevent themselves leaving muscle E, whereas if they are not strong enough to push it from either side, it will naturally remain open. Finally, if occasionally there is a

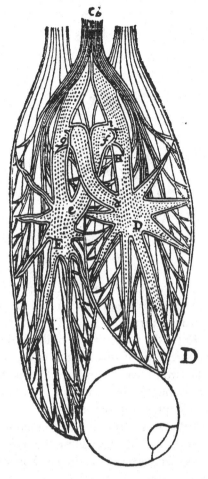

Fig. 34a

tendency for the spirits contained in the muscle D to return through *dfb*   136
or *dfe*, then the flap H can stretch and block their passage. And
similarly, between the two tubes *cg* and *dg* there is a membrane or valve,
g, corresponding to the preceding one, which naturally remains open, and
which can be closed by the spirits which come from the tube *dg*, and
opened by those coming from *cg*.

From this it can be readily appreciated that if the animal spirits in the
brain [fig. 33] tend to flow either not at all or only a little through the tubes
bf and *cg*, the two little membranes or valves *f* and *g* remain open, and thus

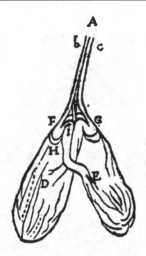

Fig. 34b

the two muscles D and E are lax and inactive, to the extent that the animal spirits that they contain pass freely from one muscle into the other, taking their course from *e* through *f* toward *d*, and reciprocally from *d* through *g* toward *e*. But if the spirits in the brain tend to enter the tubes *bf* and *cg* with some force, and if this force is equal on both sides, they immediately close the two passages *g* and *f* and inflate the two muscles D and E to the extent that they are able, in this way making them check the eye and hold it fast in its present position.

But if these spirits that come from the brain tend to run with greater force through *bf* than through *cg*, they close the little membrane *g* and open *f*, to a greater or lesser degree depending on how strongly they strike it. By these means, the spirits contained in the muscle E proceed to muscle D via channel *ef*, their speed depending on how open the membrane *f* is. As a result, the muscle D, which these spirits are unable to leave contracts, and E dilates, and so the eye is turned toward D. On the other hand, if the spirits in the brain tend to flow through *cg* with more force than through *bf*, they close the little membrane *f* and open *g*, in such a way that the spirits of the muscle D immediately return by channel *dg* into muscle E, and because of this it contracts, and turns the eye to the side.

For you will readily recognise that these spirits, being like a wind or a very fine flame, must flow promptly from one muscle to another as soon

137

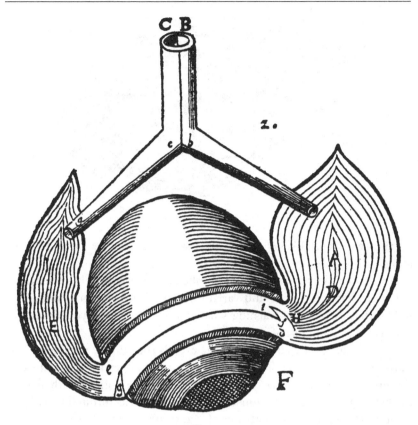

Fig. 35

as they find a passage, even though they are propelled by no other power than the inclination that they have to continue their motion in accord with the laws of nature. And you will also recognise that even though they are very mobile and very fine, they have the strength to inflate and tighten the muscles which enclose them, just as the air in a balloon hardens it and stretches the membranes that enclose it.

Now you can easily apply what I have just said about nerve A and the two muscles D and E to all other muscles and nerves, and so understand how the machine that I am describing to you can be moved in all the ways that our body can, just by the force of the animal spirits that flow from the brain into the nerves. This is because, for each motion and its contrary, you can imagine two little nerves or tubes, like *bf* and

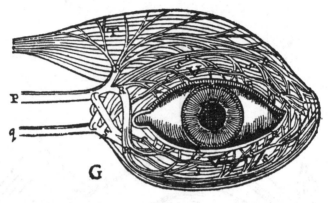

Fig. 36

cg, and two others like *dg* and *ef*, and two little doors or valves like H*fi*
138 and *g*.

And as for the ways in which the tubes are inserted in these muscles,
although there are a thousand variations, for all that it is not difficult to
judge what they are by learning what anatomy can teach you about the
external shape and use of each muscle.

If we assume, for example, that the eyelids [fig. 36] are moved by two
muscles, one of which, T, has as its sole purpose to open the upper lid and
the other, V, serves alternately to open and close both lids, then it is
easily recognised that these muscles receive the spirits through two tubes,
such as *p*R and *q*S, and that one of these tubes, *p*R, proceeds to both
muscles while the other, *q*S, proceeds only to one of them: and that the
branches R and S, being all but inserted in the same way into muscle V,
nevertheless produce two completely opposite effects there, because of
the different arrangement of their branches or fibres; and this is enough
for you to understand the others as well.

And you will have no difficulty in concluding from the foregoing that
the animal spirits are able to cause movements in all bodily parts in which
the nerves terminate, even though anatomists have failed to find any that
are visible in parts such as the eye, the heart,[22] the liver, the gall bladder,
the spleen, and so on.

---

[22] The inclusion of the heart here is peculiar, since it seems to imply that animal spirits, acting
through the nerves to the heart, cause its motion – which is in effect to say that its motion is due
to muscular action – whereas above (AT 123) Descartes has explained the motion of the heart as
being due to the fermentation of the blood.

Fig. 37

Now in order to understand specifically how this machine respires [fig. 37], imagine that it is the muscle *d* that serves to raise its chest or to lower its diaphragm, and that the muscle E is its opposite; also that the animal spirits that are in the brain cavity marked *m*, running through the pore or little channel marked *n*, which is by its nature constantly open, proceed first to the tube BF where, lowering the little membrane F, they cause those from muscle E to come and inflate muscle *d*.

Reflect next that there are certain membranes around this muscle *d*, which press on it increasingly as it is inflated, and which are arranged in such a way that, before all the spirits from muscle E have passed through

139

it, they stop in their course, and it causes them to be regorged, as it were, through the tube BF, so that those from channel n are re-directed; in this way, they proceed to cg, simultaneously forcing it open and causing the inflation of muscle E and the deflation of muscle d. And they continue to do this for as long as they endure the impetuosity of the spirits contained in muscle d, which, squeezed by the surrounding membrane, tend to be discharged from it. Then, when this impetuosity has been exhausted, they resume their course through the tube BF, so that they are unceasingly forced to inflate and deflate alternately. You should also take this to hold for the other muscles that serve the same end, and reflect that they are

140 arranged in such a way that when those such as d are inflated, the space containing the lungs is enlarged, and this causes the air to come in, just as it does when one opens a bellows; and when it is those contrary to d, this space shrinks, which causes the air to leave again.

And to understand how this machine swallows the food at the back of the mouth, assume the following. The muscle d [in fig. 37] is one of those that raise the base of its tongue and hold open the passage through which the air it is inhaling must pass in order to enter the lung; and muscle E is its antagonist and serves to close this passage. And by this means it opens that through which the food in the mouth must descend into the stomach, or rather to raise the tip of the tongue, which pushes it there; and that the animal spirits that come from the brain cavity m through the pore or tiny canal n, which by its nature remains constantly open, proceed directly into tube BF. And by these means they cause muscle d to inflate; and this muscle remains constantly inflated so long as there is no food at the back of the mouth which can press on it; but the muscle is arranged in such a way that when food is there, the spirits contained in it are immediately regorged through tube BF and cause those coming through channel n to enter into the muscle E via the tube eg, to where the spirits coming from muscle d also proceed. And thus finally the throat opens and the food descends into the stomach, and immediately

141 after this the spirits from channel n resume their flow through BF as before.

From this example, you can also understand how this machine is able to sneeze, yawn, cough, and make the motions needed to expel various excretions.

Next, in order to understand how the external objects that strike the sense organs can instigate the machine to move its members in a thousand different ways, note that the tiny fibres (which, as I have already told you, come from the innermost part of its brain and make up the marrow of the nerves[23]) are arranged in every part serving as the organ of some sense in such a way that they are easily moved by the objects of that sense.[24] And when they are moved, with however little force, they simultaneously pull on the parts of the brain from which they come and thereby open the entrances to certain pores in the internal surface of the brain. The animal spirits in the cavities of the brain immediately begin to make their way through these pores into the nerves and so into the muscles, which act so as to cause movements in the machine very like those we are naturally instigated to make when our senses are similarly affected.

Thus, for example, if fire A [fig. 38] is near foot B, the tiny parts of this fire – which as you know move very rapidly – have sufficient force to move with them the area of skin that they touch, and in this way they pull the tiny fibre *cc* which you see attached to it, and simultaneously open the entrance to the pore *de*, located opposite the point where this fibre terminates: just as when you pull on one end of a cord you cause a bell hanging at the other end to ring at the same time. 142

Now when the entrance to the pore or small tube *de* is opened this way, the animal spirits from cavity F[25] enter and are carried through it, some to the muscles that serve to pull the foot away from the fire, and some to the muscles that make the hands move and the whole body turn in order to protect itself.

But they can also be carried through the same tube *de* into many other muscles. And before pausing to explain how exactly the animal spirits follow their course through the pores of the brain and how these pores are arranged, I wish to speak to you now specifically about each of the senses that exist in this machine, and to tell you how they are related to our own.

[23] In other words, the nerves extend from the brain to various parts of the body and in doing so extend the sensory faculties based in the brain to those parts of the body.

[24] It is clear here that, on Descartes' account, the nerves have both a motor and a sensory function: he does not distinguish, as anatomists and physiologists had traditionally done, between motor nerves and sensory nerves.

[25] One could be excused for thinking, from the illustration that Clerselier supplies, that F is the pineal gland, but Descartes never refers to the pineal gland as a 'cavity'. It is almost certainly (as John Sutton has pointed out to me) one of the cerebral ventricles. It is worth noting here that the illustration to the 1662 edition does not depict F in such a way that it might so easily be mistaken for the pineal gland.

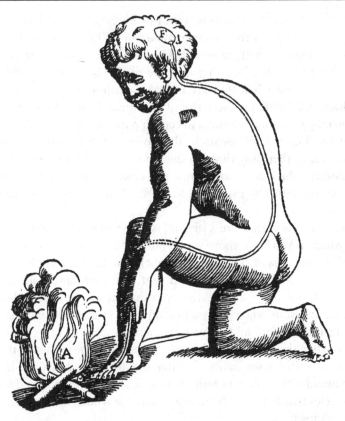

Fig. 38

[Part 3: The external senses of this machine and how they are related to ours]

Notice first that very many tiny fibres like *cc* begin to separate from one another at the internal surface of the brain where they originate, and from there they are distributed throughout the rest of the body, where they serve as the organs of the sense of touch. For although external objects do not ordinarily touch them directly, but rather touch the skin surrounding them, there is no more reason to think of the skin as the sense organ than there is to think of gloves as the sense organ when we feel something while wearing gloves.

143

And note that although the fibres I speak of are slight, for all that they extend securely all the way from the brain to the parts that are farthest away, and there is nothing in between that breaks them or which, because of pressure, hinders their activity. This is because, even though their parts are bent in countless ways, the tubes containing these fibres also carry the animal spirits to the muscles and these spirits, which always inflate the tubes to some degree, protecting the fibres from getting squashed and keeping them as taut as possible along the route from the brain, where they originate, to the places where they terminate.

Now I hold that when God unites a rational soul to this machine, as I intend to explain later on,[26] He will place its principal seat in the brain and will make its nature such that the soul will have different sensations depending on the different ways in which the nerves open the entrances to the pores in the internal surface of the brain.

Suppose for example that the tiny fibres that make up the marrow of the nerves are pulled with such a force that they are broken and separated from that part of the body to which they were attached, with the result that the structure of the machine is in some way less intact. The move-   144 ment that they will then cause in the brain, whose location must remain the same, will cause the soul to have the sensation of pain.

And if they are pulled by a force almost as great as this, but nevertheless are not broken or separated from the parts to which they are attached, they will cause a movement in the brain which, testifying to the good condition of the other parts, will cause the soul to feel a certain bodily pleasure which we call 'tingling'. And this, as you may observe, is very similar to pain as regards its cause, but quite opposite in its effect.

But if many of these tiny fibres are pulled with equal force and all together, they will cause the soul to perceive that the surface of the body touching the surface of the limb where they terminate is smooth; and if the fibres are pulled with unequal force they will cause the soul to feel it as uneven and rough.

And if they are set in motion only slightly, and separately from one another, as they are constantly by the heat that the heart transmits to other bodily parts, the soul will have no more sensation of this than of any other normal bodily function. But if this movement is increased or lessened by some unusual cause, its increase will cause the soul to have a sensation of heat, and its decrease a sensation of cold. Finally, depending on the various other ways in which they are stimulated, the fibres will

---

[26] This is presumably a reference to the projected third part of *le Monde*, which Descartes apparently never reached.

145 cause it to perceive all the other qualities that come under touch in general, such as humidity, dryness, weight, and so on.

It must be noted, however, that slight and mobile as these fibres might be, they are not sufficiently so as to be able to transmit to the brain all of the most subtle actions in nature. In fact the slightest motions that they transmit are those involving the coarser parts of terrestrious bodies. And even among those bodies there may be some whose parts, although rather coarse, can slide against the fibres so lightly that, even though they press against them or even cut through them completely, their action fails to be transmitted to the brain: just as there are certain drugs that have the power to numb or even destroy the parts to which they are applied, without causing us to have any sensation of them at all.

But the tiny fibres that make up the marrow of the nerves of the tongue, and which serve as the organ of taste in this machine, can be moved by slighter actions than those which serve for touch in general, because they are a little finer and the membranes covering them are more sensitive.

Assume, for example, that they can be moved in four different ways, by the parts of salt, acid, water, and brandy, whose sizes and shapes I have 146 already explained, and thus they can cause the soul to sense four different kinds of tastes. This they do in the following way. The parts of salt are separated from one another and are agitated by the action of the saliva, and so enter, point foremost and without bending, into the pores in the skin of the tongue. The parts of acid flow diagonally, slicing or cutting its most tender parts while giving way to the coarser ones. Those of fresh water simply glide along the top without cutting into any of its parts or advancing far into the pores. Finally, those of brandy, because they are very small, have the greatest and fastest penetration of all. From this you can easily judge how the soul will be able to sense all the other kinds of taste, if you consider in how many other ways the parts of terrestrious bodies can act against the tongue.

But what must be noted above all else here is that the parts of food which, while still in the mouth, are able to penetrate the pores of the tongue and excite the sensation of taste there, are the same as those which, while in the stomach, can pass into the blood, and from there go to join or unite with all parts of the body. And indeed only those that excite the tongue to some extent – in this way causing the soul to sense an agreeable taste – will be wholly suited for this purpose.

For just as those parts that are active to too great or too little an extent cause too sharp or too bland a taste, so also they are too piercing or too smooth to enter into the composition of the blood, or to be used for the 147 preservation of bodily parts. And there are some that are so large, or joined so tightly to one another, that they cannot be separated by the action of the saliva; or cannot in any way penetrate the pores of the tongue so as to act on the tiny nerve fibres which are used for taste there, except in so far as they may act on those that are used for the general sense of touch in other parts of the body; or do not contain pores within themselves where the small parts of the tongue, or at least those of the saliva with which it is moistened, might enter. Such as these cannot provide the soul with a sensation of taste or flavour, and hence as a rule are unsuitable to be taken into the stomach.

And this is true to such an extent that often the strength of the taste changes with the stomach's temperature, so that a food that usually seems to the soul to have an agreeable taste may in some circumstances seem bland or bitter, the reason for this being that the saliva, which comes from the stomach and always retains the qualities of the humour that predominates there, is mixed with food particles that are in the mouth, and contributes much to the way they act.

The sense of smell also depends on many tiny fibres which are projected toward the nose from the base of the brain below those two small hollowed-out parts which anatomists have compared to the nipples of a woman's breast, and which differ in no way from the nerves that serve for touch and taste, except that they do not leave the cavity of the head which 148 contains the brain as a whole, and they can be moved by even smaller terrestrious parts than those of the tongue, both because they are a little finer and because they have more direct contact with the objects that move them.

For you should also note that when this machine respires, the smallest parts of the air that enter it through the nose penetrate, by way of the pores in what is called the 'spongy' [ethmoid] bone, if not quite all the way to the cavities of the brain, at least to the space between the two membranes that enclose it, and from there they can pass out again through the palate; in the same way that when air leaves the chest it can pass into this space from the palate and leave through the nose. And on entering this space the parts of the air encounter the ends of the fibres, which are uncovered or covered only by a membrane so delicate that little force is needed to move them.

You should also note that these pores are so narrow, and so arranged, that they do not allow entry to any terrestrious parts which are coarser than those which, speaking on this subject above, I called 'odours',[27]
149 except perhaps for some of those that make up brandy, whose shape renders them very piercing.[28]

Finally, you should note that, among those very small terrestrious parts which are always found in greater abundance in air than in any other composite bodies, it is only those which are a little coarser or finer than the others, or which because of their shape are moved more or less easily, that can cause the soul to sense a variety of odours. And indeed it is only those in which these extremes are significantly moderated and mutually tempered that will cause it to have agreeable ones. For those parts that act only in the standard way cannot be sensed at all, and those that act with too much or too little force will necessarily be unpleasant to it.

As for the fibres that serve as the organ of the sense of hearing, they do not need to be as delicate as these. It is enough to think of them as being so arranged at the back of the ear cavities that they can be easily moved all together and in the same way, by the little blows with which the external air pushes against a certain very fine membrane stretched at the entrance to these cavities, and that they can be touched only by the air that lies under this membrane. For it will be these little blows that, in passing to the brain through the intermediary of these nerves, will cause the soul to conceive the idea of sound.

Note that a single one of these alone will only be able to cause a dull
150 noise which ceases in a moment, and which varies in loudness depending only on the force with which the ear is struck. But when many of them follow one another, as one sees in the vibrations of strings and of bells when they ring, then these little blows will make up one sound which the soul will judge to be smooth or harsh depending on how equal the blows are to one another, and which it will judge to be higher or lower depending on whether they follow one another slowly or quickly; so that if they follow one another a half or a third or a fourth or a fifth more quickly, they will compose a sound which the soul will judge to be higher by an octave,

[27] In his notes, La Forge notes that Descartes seems to be referring here to some other writing (or perhaps fragment missing from the present text) in which he discusses odours: see AT xi, 148 note a. Cf. *Principles of Philosophy* IV, art. 193 (AT viii, 318–19).
[28] Note that, contrary to the Galenic tradition, air does not pass through these pores.

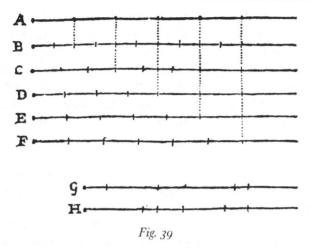

Fig. 39

a fifth, a fourth, or perhaps a major third, and so on.[29] And finally, several sounds mixed together will be harmonious or dissonant depending on the extent to which their relations are orderly, and on the extent to which the intervals between the blows making them up are equal.[30]

For example, if the divisions of the lines A, B, C, D, E, F, G, H [fig. 39] represent the little blows that make up that number of different sounds, we have no difficulty judging that those represented by the lines G and H cannot be as smooth to the ear as the others, just as rough pieces of a stone are not as smooth to touch as those of a highly polished mirror. And B must be considered to represent a sound an octave higher than A, C a fifth higher, D a fourth, E a major third, and F a full major tone. And note that A and B joined together, or ABC, or ABD, or even ABCE, are more consonant that A and F, or ACD, or ADE, and so on.[31] This seems to me to be enough 151

---

[29] Descartes is suggesting here that the overtone series, which is generated because of physical aspects of sound, is something that provides the basis for harmonic relations, a view that Mersenne had defended in detail, but which Descartes does not always adhere to: cf. Descartes to *** (Aug.–Sept. 1629), AT i. 19–20 and Descartes to Mersenne, 8 Oct. 1629, AT i. 26–7. See Gaukroger, *Descartes, An Intellectual Biography* (Oxford, 1995), 191, 286–7.

[30] This fits ill with Descartes' more functional conception of consonance set out, for example, in his letter to Mersenne cited in the last note. Fig. 39 suggests that the material on sound offered here is a barely revised version of the material in the *Compendium Musicae*. It is possible that Descartes has simply incorporated this material without much revision. It is also possible that a more functional conception would properly fit into the projected third part on the mind.

[31] Take the pitch Descartes labels 'A' as middle C. Then what Descartes is saying is that Cc, Ccg, Ccc′, and even Ccge′, are more consonant than Cf#′, Cge′, Cc′e′, and so on. That such a dissonant interval as the augmented fourth (Cf#′) should apparently be ranked with such a basic consonance as a major triad (Cge′) indicates a very mechanical approach to ranking of consonances, quite out of keeping with the far more sophisticated approach we find in the correspondence with Mersenne.

to show how the soul, when in the machine I am describing, will be able to enjoy a music that follows the same rules as ours,[32] and even how the soul will be able to make it more perfect,[33] at least when one takes into account that it is not simply the smoothest things that are most agreeable to the senses, but those that stimulate them in the most even-tempered way, just as salt and vinegar are often more agreeable to the tongue than fresh water. And this is what makes music as accommodating of thirds and sixths, and sometimes even of dissonances, as of unisons, octaves, and fifths.

There still remains the sense of vision, which I must explain a little more precisely than the others because it is more central to my subject. This sense depends in this machine, as in us, on two nerves, which must certainly be made up from many tiny fibres, as fine and as easily movable as they can be, for their role is to report to the brain the different actions of the parts of the second element, which, following what we said earlier, will enable the soul, when united with this machine, to conceive the different ideas of colours and light.

152   But because the structure of the eye also helps in accomplishing this, I must describe it here, and to make things easier I shall do so briefly, omitting many superfluous details which the curiosity of anatomists has uncovered here.

ABC [fig. 40] is a rather tough, thick membrane making up a round receptacle, as it were, in which all the other parts of the eye are contained. DEF is another, thinner membrane which is spread like a tapestry inside this. GHI is the nerve whose tiny fibres HG and GI, which spread in every direction from H toward G and I, cover the back of the eye entirely. K, L, and M are three kinds of extremely clear and transparent albumen or humours which occupy all the space in the interior of these membranes and which have, respectively, the shapes pictured here.

In the first membrane, the part BCB is transparent, and a little more arched than the rest, and rays entering it are refracted towards the

---

[32] The rules in question here are presumably harmonic, telling us what synchronic and diachronic relations between pitches are possible, and how a piece of music is to resolve. Nevertheless, when we remember that vocal music is almost exclusively at stake, how to set words to music will also be an issue, in which case questions of rhythm will also be involved. For an example of the kinds of things at issue, see the 1640 'competition' between Mersenne and Descartes' friend Johann Ban, described in ch. 6 of D. P. Walker, *Studies in Musical Science in the Late Renaissance* (Leiden, 1978).

[33] It is not clear how the soul can make music more perfect: perhaps by anticipating resolutions, filling in missing harmonies, and so on.

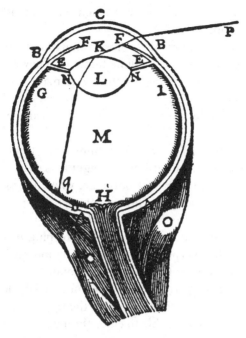

Fig. 40

perpendicular.[34] In the second membrane, the internal surface of the part EF, which faces the back of the eye, is totally black and opaque, having at its centre a small round hole called the 'pupil' which appears black in the middle of the eye when one looks at it from outside. This hole is not always of the same size, because the part EF of the membrane that the hole is in swims freely in humour K, which is very fluid, and seems to be like a little muscle that is dilated or contracted as required under the direction of the brain.

The shape of the humour marked L, which is called the 'crystalline 153 humour', is like the shape of the lenses I described in the treatise on dioptrics, whereby all the rays that come from certain points are reassembled at certain other points;[35] and its matter is less soft, or firmer,

---

[34] Towards the perpendicular, and not away from the perpendicular as we might expect, because Descartes holds the view that the speed of light is in *inverse* proportion to the optical density of the medium, with the result that it is bent toward the perpendicular on moving from air to the optically denser medium of the cornea (BCB).

[35] What Descartes shows in the *Dioptrics* is that the best shape for a convex lens to have if it is to bring all parallel rays to a single point is a hyperbola. Spherical lenses will not do this.

and consequently causes a greater refraction than that of the two other humours that surround it.

E and N are tiny black fibres that come from within the membrane DEF and completely encircle the crystalline humour. They are like so many tiny tendons which act to change its shape and make it a little flatter or more arched as needed. Finally, o, o are six or seven muscles attached to the eye from the outside, which can move it very easily and very quickly in all directions.

Now the membrane BCB, and the three humours K, L, and M, being very clear and transparent, do not prevent the light rays entering through the pupil from penetrating to the back of the eye where the nerve is located, nor from acting on it as easily as if it were completely exposed. They serve to protect it from harm from the air and other external bodies that could easily injure it if they touched it, as well as keeping it so delicate and sensitive that it is not surprising that it can be moved by such barely perceptible actions as those I take here to be colours.

The curvature of the part of the first membrane labelled BCB and the
154 refraction that occurs there is what makes the rays from objects located towards the sides of the eyes able to enter through the pupil. They thus enable the soul, without the eye moving, to see a larger number of objects than it otherwise could. If for example the ray PBKq did not bend at point B it would not be able to pass between the points F, F and so reach the nerve.

The refraction that occurs in the crystalline humour serves to make vision stronger and at the same time more distinct. For you should note that the shape of this humour is such that, considering refractions that occur in other parts of the eye and the distance of objects, when the vision is trained on a particular point, it causes all the rays that come from this point and enter the eye through the pupil to collect together again at another point at the back of the eye. They collect together at exactly one part of the nerve located there; and in the same way, other rays entering the eye are prevented from touching the same part of this nerve.

For example, when the eye is so arranged as to look at point R [fig. 41], the crystalline humour is disposed in such a way as to make all the rays RNS, RLS, and so on, collect together again exactly at point S, and by the same means prevents any of those coming from points T and X etc. from arriving there. For it assembles all those from point T around point V, and
155 those from point X around point Y, and so on for the others. Whereas if there were no refraction in this eye, the object R would send only one of

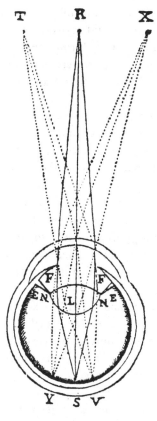

Fig. 41

its rays to point s, the others being spread here and there throughout the area VY; and in the same way the points T and X, and all those in between, would each send its rays towards this same point s.

Now it is sufficiently obvious that object R must act more strongly on the part of the nerve at point s when it sends a large number of rays there than when it only sends one, and that part s of the nerve is going to relay the action of object R to the brain more distinctly and faithfully when it receives rays from object R alone and none from any of the various other objects.

The black colour of the internal surface of the membrane EF[36] and of

---

[36] EF is the iris.

the tiny fibres EN[37] also helps in making vision more distinct. This is because, in accord with what I said above on the nature of colour,[38] it deadens the force of the rays reflected from the back of the eye toward the front, thereby preventing them from returning to the back, where they might produce confusion. For example, if F and N were not black, the rays from object X, on reaching the point Y of the nerve, which is white, would be reflected from there in every direction toward N and F, from where they could be turned back again toward S and V and there interfere with the action of those coming from points R and T.

156   The change of shape that occurs in the crystalline humour allows objects lying at different distances to paint their images distinctly on the back of the eye. For, following what has already been said in the treatise on dioptrics, if the humour LN, for example [fig. 42], is shaped in such a way that it causes all the rays from point R to strike the nerve exactly at point S, this same humour cannot also make the rays from point T, which is closer, or those from point X, which is further away, collect there too without its shape being changed. It will make TL go toward H and TN toward G; and on the other hand it will make XL go toward G, and XN toward H, and so with the others. So that in order to represent point X distinctly, the whole shape of this humour LN has to be changed and become slightly flatter, like that marked I; and to represent point T it has to become slightly more arched, like that marked F.

    The changes in the size of the pupil serve to moderate the strength of vision; for when the light is too bright it needs to be smaller, so that the rays of light that enter the eye are not so numerous that they damage the nerve; and it has to be larger when the light is weak, so that enough of the rays enter to be sensed. In addition, in the case where the light remains constant, it is necessary that the pupil be larger when the object viewed is distant than when it is near; for if only as many rays from point R enter the pupil of eye 7 [fig. 43] as are needed in order to be sensed, for example, the same number must enter eye 8, and in consequence its pupil must be larger.

157   The small size of the pupil also serves to make vision more distinct; for you should note that, no matter what shape the crystalline humour may have, it cannot make rays that come from different points of the object

---

[37] EN is the ciliary body.
[38] This may be a reference to the account that will later be published in the *Meteors*: the relevant section is translated above as Appendix 2.

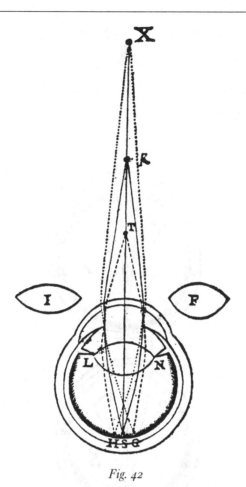

Fig. 42

collect together exactly at correspondingly different points. Rather, if the rays from point R [see fig. 41 above], for example, come together at point S then, of the rays from T, only those that pass through the circumference and the centre of one of those circles that can be described on the surface of this crystalline humour can be collected exactly at point V, and the others – which will be fewer the smaller the pupil – are consequently going to touch the nerve at other points, and cannot fail to cause confusion there. So that if the vision of an eye is less strong at one time than at another, it will also be less distinct, whether this arises from the distance of the object or the weakness of the light. This is because

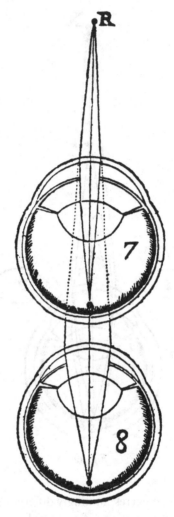

Fig. 43

the pupil's being larger when the light is less strong makes vision more confused.

From this it also follows that, at any one time, the soul will never be able to see more than a single point on the object very distinctly, namely that on which all parts of the eye are trained at that time, and the others will appear much more confused the farther away from this one is, because if rays from point R, for example, all collect exactly at point s, those from point x will collect even less exactly at y than those from point t collect at v; and we must conclude that the same is true of others in proportion to their distance from r. But the muscles o, o [see fig. 40 above], 158 turning the eye very quickly in every direction, serve to remedy this defect, for in no time at all they can successively turn the eye to all points on the object, thus letting it see all points distinctly one after the other.

I shall not add here the details of what makes it possible for the soul to conceive of all the differences in colour, since I have already dealt with that.[39] Nor shall I say what objects of vision must be agreeable or disagreeable to it; for from what I have already said about the other senses, it is easy to grasp that light that is too strong will injure the eyes and moderate light must refresh them; and that, amongst the colours, green, which consists in the most moderate action (which by analogy one can speak of as the ratio 1:2), is like the octave among musical consonances, or like bread among the foods that one eats, that is, it is the most universally agreeable.[40] And finally, all the different fashionable colours which are more refreshing than green are like the chords and passages of a new tune, played by an excellent lutenist, or the stews of a good cook, which stimulate the sense and first make it feel pleasure but then become tedious more quickly than simple and ordinary objects.

It remains only for me to tell you what it is that will give the soul the 159 means to sense position, shape, distance, size, and other similar qualities which are not related to one sense like those we have spoken of up to now, but are common to touch and vision and even in some respects to the other senses.

Note first that if hand A touches body c [fig. 44], for example, the parts of the brain B from which the tiny fibres of its nerves issue, will come to be arranged differently from how they would have been had it touched a body of different shape, or size, or location. These, then, are the means by which the soul will be able to tell the shape of a body, as well as its shape, size, and all other similar qualities. Similarly, if the eye D is turned toward object E [fig. 45], the soul will be able to tell the position of this

[39] Actually, as we have seen, Descartes has not dealt with this in the present treatise and is probably referring to the account that will subsequently appear in the *Meteors*. See above, pp. 85–96.

[40] What the basis of this remark is is unclear, and although various writers have made suggestions about the relations between colours and sounds, the attempt to quantify green on a par with an octave certainly cannot be sustained. It is worth noting that Descartes will later advise Elizabeth to rid her mind of sad thoughts by reflecting on the greenness of a wood (Descartes to Elizabeth, May/June 1645, AT iv. 220).

Fig. 44

object, inasmuch as the nerves from this eye are disposed in a different way than they would be if it were turned toward some other object. And it will be able to tell its shape, inasmuch as rays from point 1 collecting on the nerve called the optic nerve at point 2, and from point 3 at point 4, and so on, will trace a shape there which corresponds exactly to the shape of E. And it will be able to tell what distance it is from point 1 for example, inasmuch as the shape of the crystalline humour will be different – in order to make all the rays from this point collect at the back of the eye exactly at point 2, which I assume to be in the middle – than if it were 160 nearer or farther, as I have already said. And moreover it can tell the distance of point 3, and all others whose rays enter at the same time, because the crystalline humour will be so disposed that the rays from point 3 will not collect so exactly at point 4 as will those from point 1 at

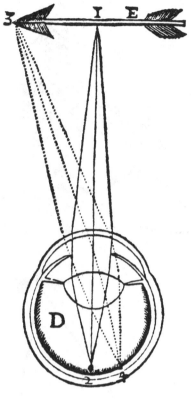

Fig. 45

point 2, and similarly with the others, and their action will be proportionately weaker, as has already been said. And finally, the soul will be able to tell the size and all other similar qualities of visible objects simply through its knowledge of the distance and position of all their points, just as, conversely, it will sometimes judge their distance from the opinion it has of their size.

Notice also that if two hands *f* and *g* [fig. 46] each hold sticks *i* and *h* with which they touch the object к, then even though the soul is otherwise ignorant of the length of the sticks, nevertheless, because it can tell the distance between the points *f* and *g*, and the sizes of the angles *fgh* and *gfi*, it will be able to tell, as if by a natural geometry, where the object к

---

[41] That is to say, if we know the length of the base (*fg*) and the size of the base angles (*fgh* and *gfi*) we can calculate the length of the sides and the distance to к.

Fig. 46

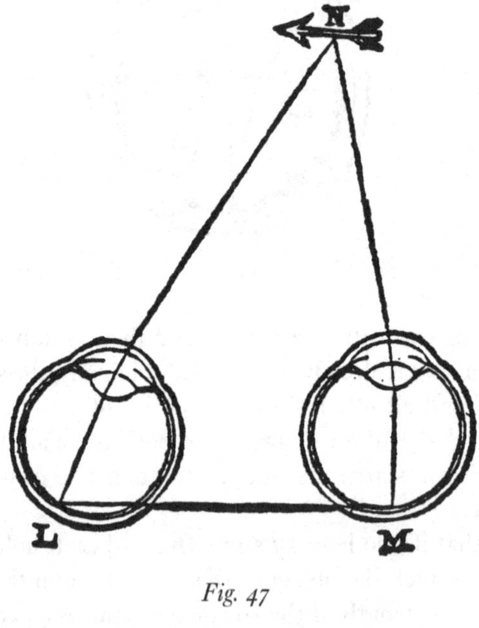

Fig. 47

is.[41] And in just the same way, if the two eyes L and M [fig. 47] are turned towards the object N, the size of the line LM and of the two angles LMN and MLN will tell it where point N is.

But it can often happen that the soul is mistaken in this. For suppose

134

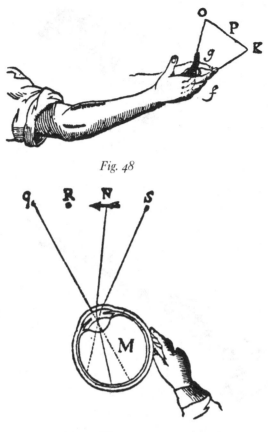

Fig. 48

Fig. 49

first that the position of the hand, or that of the eye, or the finger, is constrained by some external cause: then its position will not correspond so
exactly with that of the tiny parts of the brain where the nerves originate 161
than if it depended on the muscles alone. And so the soul, which will
sense this only through the mediation of the parts of the brain, must be
mistaken in that case.

Suppose, for example, that hand *f* [fig. 48], being itself disposed to turn
toward o, finds itself constrained by some external force to remain turned
toward κ. Then the parts of the brain from which the nerves originate will
not be arranged in just the same way as they would if the hand were
turned toward κ by the force of the muscles alone. Nor will they be
arranged as they would if the hand were really turned towards o. Rather,

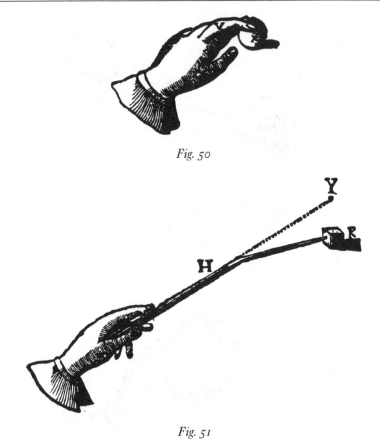

Fig. 50

Fig. 51

they are arranged in a manner intermediate between these two, that is, as if turned toward P. And so the arrangement that this constraint imposes on the tiny parts of the brain will cause the soul to judge that the object K is at point P, and that it is a different object from that which is touched by the hand *g*.

Similarly, if eye M [fig. 49] is turned away from object N by force, and disposed as if looking toward *q*, the soul will judge that the eye is turned toward R. And because in this case the rays from object N will enter the eye in the way that those from s would if the eye were really turned toward R, it will believe that this object N is at point s, and that it is a different object from that which the other eye is looking at.

Similarly, the two fingers *t* and *v* [fig. 50], touching the little ball x, will

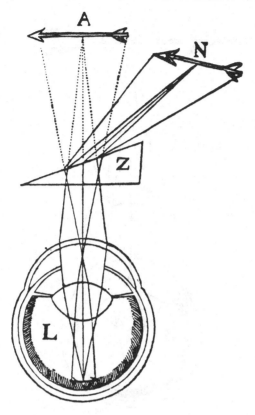

Fig. 52

cause the soul to judge that they are touching two different things because they are crossed and kept forcibly from their natural position.[42]

Moreover, if the rays – or any other lines along which the actions of distant objects pass toward the senses – are curved, the soul, which generally supposes them to be straight, may be deceived. For example, if the stick HY [fig. 51] is curved towards K, it will seem to the soul that object K which the stick touches is in the direction of Y. And if eye L [fig. 52] receives rays from object N through the glass Z, which bends them, it will seem to the soul that this object is in the direction of A. Similarly, if eye B [fig. 53] receives rays from point D through glass c, 162

---

[42] The tactile illusion caused by crossed fingers had been reported on in antiquity, e.g. in Aristotle, *Parva naturalia* 460b20. Descartes repeats the example in Discourse 6 of the *Dioptrics* (AT vi. 141–2).

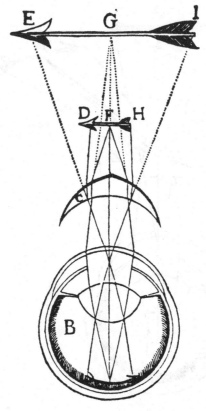

Fig. 53

which I am assuming bends them as though they were coming from point E, and bends those from F as though they were coming from point G, and so on with the others, it will appear to the soul that the object DFH is as distant and as large as EGI appears to be.

And, in conclusion, we must note that none of the means by which the soul tells distance will be completely certain, for the following reasons. For with angles like LMN and MLN [see fig. 47] and so on, the change in these is insignificant when the objects are more than fifteen or twenty feet away. And changes in the shape of the crystalline humour are even more insignificant than these for objects more than two or three feet from the eye. Finally, perspectival techniques readily show us how easy it is to be deceived when we judge the distances of objects on the basis of our opinion of their size, or from the fact that rays from different points of an object do not come together in exactly the same way at the back of the eye. For when they have shapes such that they are smaller than we imagine 163 them to be, and their colours make them somewhat obscure, and their outlines somewhat indefinite, then all these cause them to appear to be more distant and larger than they are.

Now, having explained for you the five external senses, as they are in this machine, I must also tell you something about the internal senses it contains.

[Part 4: On the internal senses which are to be found in this machine]

When the fluids which, as I mentioned earlier, act as a sort of *aqua fortis* in the stomach, passing unceasingly from the whole mass of the blood through the whole length of the arteries, do not find sufficient food to dissolve there to use up all their force, they act on the stomach itself in a particular way and, agitating the tiny fibres of its nerves more strongly than usual, the parts of the brain from which they originate are moved. This will cause the soul, when it is united to this machine, to conceive the general idea of hunger. And if these fluids are so disposed as to act against certain particular foods rather than others, much as ordinary *aqua fortis* dissolves metals more easily than wax, they will also act in a particular way against the nerves of the stomach. This will cause the soul at such times to conceive an appetite to eat certain foods rather than others.[43] Now 164 these fluids are gathered together mainly at the bottom of the stomach, and it is there that they cause the sensation of hunger.

But many of their parts also rise continually towards the throat, and when they do not come there in sufficient amounts to moisten it and fill its pores in the form of water,[44] they rise instead in the form of air or smoke, and then they act against the nerves in an unusual fashion, causing a movement in the brain that will make the soul conceive of thirst.

---

[43] At this point a Latin sentence appears in the text which its first editor, La Forge, considered to have been a marginal comment by a reader which a careless copyist had incorporated into the French text. The sentence reads: 'Here one can note the admirable structure of this machine, which is such that hunger results from fasting. For the blood becomes more acrid while it is circulating, and thus the liquid that goes from the blood into the stomach agitates the nerves there more strongly than is usual, and it does so in a particular way if the constitution of the blood also happens to be peculiar. And from this derive the longings of [pregnant] women.'

[44] This is an account of the formation of saliva. See Descartes to Mersenne, 30 July 1640, where a different account of the production of saliva is offered.

Similarly, when the blood which enters the heart is purer and more subtle, and is kindled there more easily than usual, this disposes the tiny
165 nerve there in the way required to cause the sensation of joy; and in the way required to cause the sensation of sadness when this blood has the opposite qualities.

And from this it is easy to grasp what there is in this machine that corresponds to all the other internal sensations in us; whence it is time for me to begin explaining how the animal spirits make their way through the cavities and pores of its brain, and which functions depend on them.

If you have ever had the curiosity to look closely at the organs in our churches, you will know how the bellows push the air into certain receptacles, which for this reason are named wind chests; and also how this air passes from there into one or another of the pipes, according to the different ways in which the organist moves his fingers on the keyboard. You can think of our machine's heart and arteries, which push the animal spirits into the cavities of the brain, as being like the bellows of an organ, which push air into the wind chests; and of external objects, which displace certain nerves, causing spirits from the brain cavities to enter certain pores, as being like the fingers of the organist, which press certain keys and cause the wind to pass from the wind chests into certain pipes. And just as the harmony of organs depends not on the externally visible arrangement of pipes or on the shape of the wind chests or other parts but
166 solely on three factors, namely the air that comes from the bellows, the pipes that make the sound, and the distribution of air in the pipes; so too, I would point out, the functions that we are concerned with here do not depend at all on the external shape of the visible parts which the anatomists distinguish in the substance of the brain and in its cavities, but solely on three factors, namely, the spirits that come from the heart, the pores of the brain through which they pass, and the way in which the spirits are distributed in these pores. Thus my sole task here will be to explain to you, in a systematic way, what is most important in these three.

First, as to animal spirits, they can be more or less abundant, and their parts more or less coarse, and more or less agitated, and more or less uniform, from one time to another. And it is by virtue of these four differences that all of the various humours or natural inclinations in us (at least in so far as these do not depend on the constitution of the brain, or on particular affections of the soul) are also represented in this machine. For if these spirits are exceptionally abundant, they are able to excite in it movements similar to those that testify in us to generosity, liberality, and love; confidence and courage if their parts are very strong and coarse, and of constancy if their parts are also more equal in shape, force, and size; promptitude, diligence, and desire if they are exceptionally agitated; and tranquillity of spirit if their agitation is exceptionally uniform. Whereas if the same qualities are lacking, on the other hand, these same spirits are able to excite movements in it just like the movements in us that testify to malice, timidity, inconstancy, tardiness, and ruthlessness. 167

And note that all the other humours or natural inclinations depend on those just mentioned. Thus the joyous humour is made up from promptitude and tranquillity of spirit, and generosity and confidence serve to make the joyous humour more perfect. The sad humour is made up from tardiness and restlessness, and malice and confidence can augment it. The choleric humour is made up of promptitude and restlessness, and malice and confidence strengthen it. Finally, as I have just said, liberality, generosity, and love depend upon an abundance of spirits, and form in us that humour which renders us obliging and benevolent to everyone. Curiosity and the other inclinations depend on the agitation of their parts, and similarly with the others.

But because these same humours, or at least the passions to which they dispose us, also depend to a great extent on the impressions made in the substance of the brain, you will be able to understand them better later on, and I shall restrict myself here to telling you the causes of the differences in the spirits.

When the juice from the food that passes from the stomach into the veins is mixed with the blood, it always communicates some of its own qualities to it and, among other things, it usually makes it coarser when they are freshly mixed: so that then the tiny parts of this blood which the heart sends to the brain, in order to make the animal spirits there, are usually neither particularly agitated, or strong, or abundant. Consequently they do not as a rule make this machine as quick or lively as it becomes some time after digestion is complete, and after the same blood, having passed and repassed through the heart several times, has become finer. 168

The air of respiration is also mixed in some way with the blood before it enters the left cavity of the heart, making the blood kindle more strongly and producing livelier and more agitated spirits there in dry weather than in humid weather, just we observe all kinds of flame to be more ardent at such times.

When the liver is well disposed and transforms completely the blood
169 that must enter the heart, the spirits that issue from this blood are
correspondingly more abundant and more uniformly agitated. And
should the liver be compressed by its nerves, the subtlest parts of the
blood that it contains will, by rising straight to the heart, produce spirits
correspondingly more abundant and livelier than is usual, but their
agitation will not be so uniform.

If the gall-bladder, which is designed to purge the blood of those of its
parts most suited to being kindled in the heart, fails in its task, or if its
nerve acts to retract it and the matter contained in it is regorged into the
veins, then the spirits there will be all the more lively and unequally
agitated.

On the other hand, if the spleen, which is designed to purge the blood
of those of its parts least suited to being kindled in the heart, is ill
disposed, or if, squeezed along by its nerves or by any other body at all,
the matter contained in it is regorged into the veins, then the spirits will
be all the less abundant, lively, and uniformly agitated.

In short, whatever can cause a change in the blood can also cause one
in the spirits. But above all, the little nerve that terminates in the heart,
by virtue of being able to dilate and contract the two entrances through
which the venous blood and the pulmonary air descend, as well as the two
exits through which the blood is exhaled and driven into the arteries, can
cause a thousand differences in the nature of the spirits: just as the heat
of certain enclosed lamps used by alchemists can be moderated in many
ways, depending on the degree to which one opens now the passage
170 through which the oil or other fuel for the flame comes in, and now that
by which the smoke goes out.

*[Part 5: On the structure of the brain of this machine, and how the spirits
are distributed there so as to cause its movements and its sensations]*

Secondly, as far as the pores of the brain are concerned, they must be
thought of as being no different from the spaces that exist between the
fibres of a tissue;[45] for the brain is in fact just a tissue constituted in a
particular way, as I shall now try to explain to you.

---

[45] As Hall, *René Descartes: Treatise of Man*, points out in the notes to his translation (p. 77 n.123),
Descartes' term '*tissu*' means something textured in some way, and is not confined to tissues in the
modern sense.

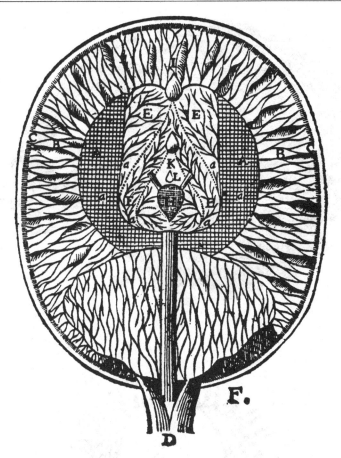

Fig. 54

Consider its surface AA [figs. 54 and 55], which faces cavities EE, to be a somewhat dense, compact net or mesh all of whose links are so many tiny tubes through which the animal spirits can enter and which, since they always face gland H from where these spirits originate,[46] can easily turn this way and that toward the different points on this gland: as you 171 can see from the different ways in which they turn in forms 48 and 49 [right and left sides respectively of fig. 56]. And suppose that from each part of this net there arise several extremely fine fibres, some of which are generally longer than others; and that once these fibres have been

[46] Gland H is the pineal gland.

143

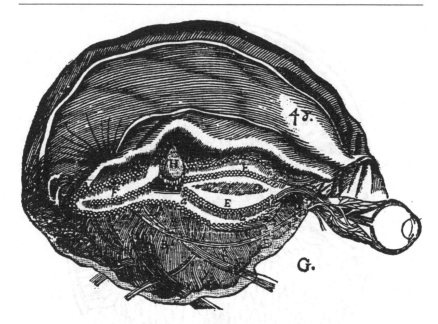

Fig. 55

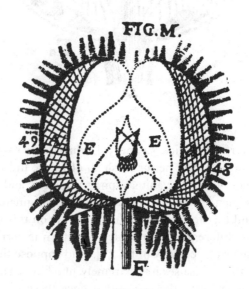

Fig. 56

variously interlaced throughout the region marked B [see fig. 55], the longer ones descend towards D and from there make up the marrow of the bones, and go on to spread through all the parts of the body.

Assume also that the chief qualities of these tiny fibres are the ability to be flexed readily in every way simply by the force of the spirits that strike them, and the ability to retain, as if made of lead or wax, the last flexure received until a contrary force is applied to them.

Finally, assume that the pores in question are just the gaps between these fibres, and that they can be variously dilated and contracted by the force of the spirits entering them, depending on the strength of this force and on how abundant they are; and that the shortest of these fibres leads into space *cc* [see fig. 54], where each of them terminates at the end of one of the tiny vessels there which is nourished by it.

Third, if I am to explain all the particular features of this tissue most easily, I must now begin to tell you about the distribution of these spirits.

The spirits never stop for a single moment in any one place; but as fast 172 as they enter the brain cavities EE through the holes in the little gland marked H, they tend first in the direction of the tubes *aa* which are most directly facing them; and if these tubes *aa* are not open wide enough to take them all in they at least take in the strongest and liveliest of the parts, while those which are feeblest and most superfluous are driven towards the passages I, K, and L, which face the nostrils and the palate. That is to say, the most agitated are driven towards I, which – when they still have a lot of force and when the passage is not sufficiently clear – they sometimes leave with so much violence that they tickle the internal parts of the nose, which causes sneezing; and then the others are driven towards K and L, from where they can leave very easily because the passages there are very large. If they fail to do this they are driven back toward tubes *aa* in the inside surface of the brain, and they promptly cause a dizziness or vertigo which interferes with the functioning of the imagination.

And note in passing that the more feeble parts of the spirits derive less from the arteries embedded in the gland H than from those which divide into a thousand very tiny branches and carpet the cavities of the brain. Note also that they can also thicken into phlegm, but they never do this in the brain, except in the case of grave illness; ordinarily this takes place in those large spaces beneath the base of the brain between the nostrils and the gullet, just as smoke is readily transformed into soot in the 173 chimney flue, but never in the hearth where the fire is.

Note also that when I say that spirits, in issuing from the gland H, tend toward those places on the inside surface of the brain which are the most directly opposite, I do not mean that they always tend in the direction facing them in a straight line, but only in the direction in which the disposition of the brain makes them tend.

Now, the substance of the brain being soft and pliant, its cavities would be very narrow and almost all closed – as they appear in the brain of a dead man – if no spirits entered them. But the source that produces these spirits is usually so abundant that in entering these cavities they have enough force to push the matter that surrounds them outward in all directions, causing it to expand and tighten all the tiny nerve fibres coming from it, in the same way that a moderate wind can fill the sails of a ship and tighten all the ropes to which they are attached. It follows that at such times this machine, being disposed to respond to all the actions of the spirits, represents the body of a man who is *awake*. Or at least the spirits have enough force to push against and stretch some parts while the others remain free and relaxed, as happens in various parts of the sail when the wind is too feeble to fill it. And at such times this machine represents the body of a man who is asleep and who has various dreams as he sleeps. Imagine for example that the difference between the two 174 figures M and N [figs. 57, 58, and 59] is the same as that between the brain of a man who is awake and that of a man who is asleep, and who dreams while sleeping.

But before I speak in greater detail about sleep and dreams, I ask you to consider what is most noteworthy about the brain during the time of waking: namely, how ideas of objects are formed in the place assigned to the imagination and to the common sense, how these ideas are retained in the memory, and how they cause the movement of all the bodily parts.

You can see in the diagram marked M [fig. 58] that the spirits that issue from gland H, having dilated the part of the brain marked A, and having partly opened all the pores, flow from there to B, then to C, and finally into D, from where they stream out into all the nerves. And in this way they keep all the tiny fibres that compose the nerves and the brain so taut that even those actions that have barely enough force to move them are easily communicated from one end to the other, and the detours through which they pass do not hinder this.

But in case these detours prevent you from seeing clearly how the ideas of objects that strike the senses are formed, note in this illustration

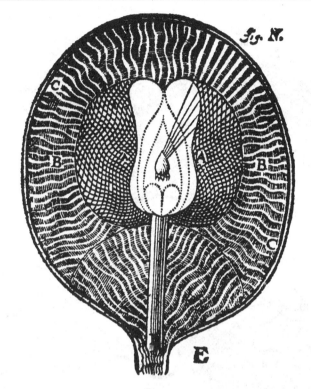

Fig. 57

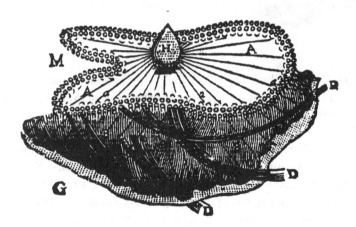

Fig. 58

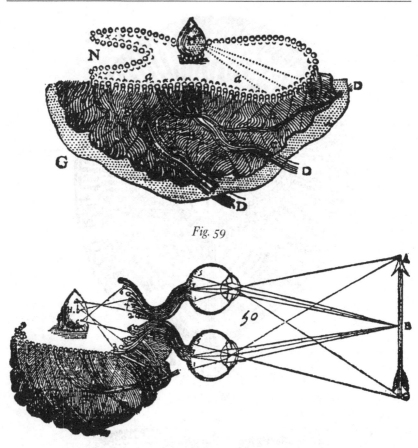

Fig. 59

Fig. 60

175 [fig. 60] the tiny fibres 12, 34, 56, and so on, which constitute the optic nerve and extend from the back of the eye at 1, 3, 5 to the inside surface of the brain at 2, 4, 6. Now assume that these fibres are so arranged that if the rays coming from point A on the object, for example, happen to press on the back of the eye at point 1, they pull the whole of fibre 12 and enlarge the opening of the tiny tube marked 2. In the same way, the rays coming from point B enlarge the opening of the tiny tube 4, and so with the others. Thus, owing to the different ways in which the rays exert pressure on the points 1, 3, and 5, to trace a figure on the back of the eye corresponding to that of object ABC, as we have already said, it is evident that the different ways in which the tiny tubes 2, 4, 6, and so on are opened

148

by the fibres 12, 34, 56 etc., must also trace it on the inside surface of the brain.

Suppose next that the spirits that tend to enter each of the tiny tubes 2, 4, 6 and so on do not come indifferently from all points on the surface of gland H but each from one particular point: those coming from point *a* on this surface for example tend to enter tube 2, and those from points *b* and *c* tend to enter tubes 4 and 6, and so on with the others. As a result, at the same instant that the openings to these tubes enlarge, the spirits 176 begin to issue from the corresponding points on the gland more freely and more rapidly than they otherwise would. Thus, just as the figure corresponding to that of the object ABC is traced on the inside surface of the brain depending on the different ways in which tubes 2, 4, 6 are opened, so that figure is traced on the surface of the gland depending on the ways in which the spirits issue from points *a*, *b*, and *c*.

And note that by figure I mean not only things that somehow represent the position of the edges and surfaces of objects, but also anything which, as I said above, can give the soul occasion to sense movement, size, distance, colours, sounds, smells, and other such qualities; and even things that can make it sense pleasure,[47] pain, hunger, thirst, joy, sadness, and other such passions. For it is easy to understand that tube 2, for example, will be opened differently as the action causing it differs, whether this action is that causing sensory perception of the colour red, or of pleasure, or the action that I said causes sensory perception of the colour white, or of pain; and the spirits that issue from point *a* will tend to move toward this tube in a different way depending on the differences in its manner of opening, and likewise for the others.

Now among these figures, it is not those imprinted on the organs of external sense, or on the inside surface of the brain, that should be taken as ideas, but only those traced in the spirits on the surface of gland H, where the seat of the imagination and the common sense is. That is to say, 177 only these should be taken as the forms or images which, when united to this machine, the rational soul will consider directly when it imagines some object or senses it.

And note I say 'imagine' or 'sense'. For I wish to apply the term 'idea' generally to all the impressions which the spirits are able to receive as

[47] The term Descartes uses here – 'chatoüillement' – literally means tickling, but tickling is not a passion, and what he seems to be referring to is a kind of light-hearted sensory pleasure. In the next sentence it is contrasted with pain.

they issue from gland H.[48] And when these depend on the presence of objects they can all be attributed to the common sense; but they may also proceed from other causes, as I shall explain later, and they should then be attributed to the imagination.

And I could add something here about how the traces of these ideas pass through the arteries to the heart, and thus radiate throughout the blood; and about how they can sometimes even be caused by certain actions of the mother to be imprinted on the limbs of the child being formed in her womb.[49] But I shall content myself with telling you more about how the traces are imprinted on the internal part of the brain, marked B, which is the seat of memory.

To this end, imagine that after issuing from gland H spirits pass through tubes 2, 4, 6 and the like, into the pores or gaps lying between the tiny fibres making up part B of the brain. And suppose that the spirits are strong enough to enlarge these gaps a little, and to bend and arrange any fibres they encounter in various ways, depending on the different ways in which the spirits are moving and the different openings of the tubes into which they pass. And they do this in such a way that they also trace figures in these gaps, corresponding to those of the objects. At first they do this less easily and perfectly here than on gland H, but they gradually improve as their action becomes stronger and lasts longer, or is repeated more often. Which is why in such cases these patterns are no longer easily erased, but are preserved in such a way that the ideas that were previously on this gland can be formed again long afterwards without requiring the presence of the objects to which they correspond. And this is what memory consists in.

For example, when the action of the object ABC, by increasing the degree to which tubes 2, 4, and 6 are open, causes the spirits to enter into them in greater quantity that they would otherwise, it also causes them, as they pass further on towards N,[50] to have the force to form certain

---

[48] This understanding of ideas as something corporeal can be found throughout Descartes' writings, especially in writings such as this and the *Rules*, where he is concerned with the psycho-physiology of perceptual cognition.

[49] Cf. Discourse 5 of the *Dioptrics*: 'I could go even further and show you how the picture can sometimes pass from [the pineal gland] through the arteries of a pregnant woman, to some particular part of the infant that she carries in her womb, forming there those birthmarks that cause learned men to marvel' (AT vi. 159). The idea that the physiognomy of the child can be affected by the images in the mother's mind was widespread at this time. Descartes is only undertaking to explain the phenomenon (or tell us that he can explain it).

[50] It is not at all clear what 'N' Descartes is referring to here: perhaps that in fig. 59.

Fig. 61

passageways there, which remain open even after the action of the object ABC has ceased; or at least, if they close up again, they leave a particular arrangement in the fibres composing this part of the brain N, and by these means they can be opened more easily later than if they had not been opened previously. Similarly, if one were to pass several needles or engraver's points through a linen cloth as you see in the cloth marked A [fig. 61], the tiny holes that would be made there would stay open, as at *a* and *b*, after the needles are withdrawn; or if they did close again, they would leave traces in the cloth, as at *c* and *d*, which would make them very easy to open again. 179

Similarly, it must be noted that if one were to re-open just some of them, like *a* and *b*, this in itself would cause others such as *c* and *d* to re-open at the same time, especially if they had all been opened together several times and had not usually been opened separately. This shows how the recollection of one thing can be excited by that of another which had been imprinted in the memory at the same time. For example, if I see two eyes with a nose, I immediately imagine a forehead and a mouth, and all the other parts of a face, because I am unaccustomed to seeing the

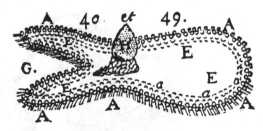

Fig. 62

former without the latter. And seeing fire, I am reminded of heat, because I have felt this in the past when seeing fire.

Consider also that gland H is composed of very soft matter which is not joined to or part of the substance of the brain but attached only to certain little arteries whose membranes are somewhat relaxed and pliant, and that it is kept in balance as it were by the flow of blood which the heat of the heart drives in its direction; so that very little is required to make it incline or lean, whether a little or a great deal, whether to this side or to that, and so to make the spirits that issue from it proceed to particular regions of the brain rather than others.

180   Now there are two main causes – aside from the force of the soul, which I shall deal with later – of the gland's moving in this way, which I shall set out here.

First, there are the differences among the tiny parts of the spirits that issue from it. For if these spirits all had exactly the same force and if there were no other cause determining that the gland lean this way or that, then they would flow equally in all its pores and keep it erect and immobile at the centre of the head, as is represented in [fig. 62]. But just as a body attached only by threads and kept in the air by the force of the smoke issuing from a furnace would float here and there incessantly, as the different parts of the smoke acted in different ways against it, so the tiny parts of the spirits that hold this gland and keep it in its place almost always differ among themselves in some way, and they must agitate it and make it lean now to one side and now to another. Thus we can see [in fig. 63] that not only is the centre of gland H a slight distance from the centre of the brain, marked *o*, but also that the ends of the arteries holding it up are curved in such a way that nearly all the spirits that the arteries bring to it proceed through that region of its surface in the direction of the tiny tubes 2, 4, and 6, and in this way they open the

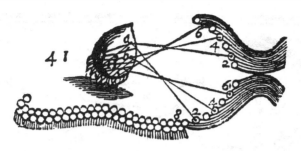

Fig. 63

pores facing in that direction to a much greater extent than they do other pores.

The main effect of this is that the spirits, issuing specifically from 181 certain regions on the surface of this gland and not from others, have the force to turn the tiny tubes from the inside surface of the brain, into which they flow, towards those places from where they issue, unless they are already pointed in that direction; and by these means they are able to make the bodily parts to which these tiny tubes correspond turn towards the places matching these regions on the surface of gland H. And note that the idea of this movement of bodily parts just consists in the way in which the spirits flow from the gland, and thus it is its idea that is the cause of the movement.

Here [fig. 64] for example, we can assume that what makes tube 8 turn towards point *b* rather than toward some other point is simply that the spirits that issue from this point tend toward it with a greater force than do any others. And the same thing will cause the soul to sense that the arm is turned toward object B, if it is already in this machine, as I shall later suppose it to be. For we must imagine that all the points of the gland toward which tube 8 can be turned correspond to places toward which the arm marked 7 can be turned, so that what makes the arm turn toward object B now is simply that this tube is facing point *b* of the gland. But if the spirits, changing their course, turn these tubes toward some other point on the gland, say toward *c*, then the tiny fibres 8 and 7, which emerge nearby and proceed to the muscle of this arm, in changing their position by the same means, would close up certain pores of the brain 182 near D, enlarging others. This would make the spirits, which would thereby pass into these muscles in a different way from that they do now, promptly turn this arm toward object C. Reciprocally, if some action

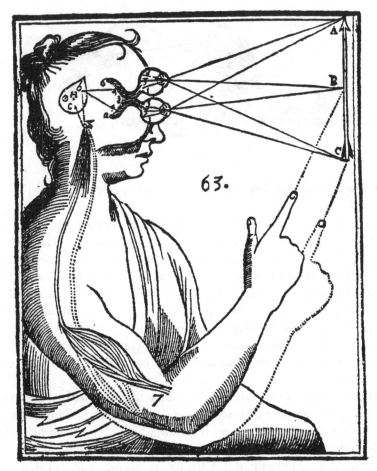

Fig. 64

other than that of the spirits which enter through tube 8 were to turn this same arm toward B or C, this would make this tube 8 turn toward points *b* or *c* of the gland. As a consequence, an idea of this movement would be formed at the same time, at least if one's attention were not diverted, that is to say, if gland H were not prevented from leaning toward 8 by some different, stronger action. Thus in general we should take it that each tiny tube on the inside surface of the brain corresponds to a bodily part, and that each point on the surface of gland H corresponds to a direction in which these parts can be turned: in this way, the movements of these parts and the ideas of them can cause one another in a reciprocal fashion.

Moreover, occasionally, when the two eyes of this machine, as well as the organs of several of the other senses, are directed towards one and the same object, there are formed in the brain not several ideas of it but only one. To understand how this can be, we must assume that spirits leaving the same points on the surface of the gland H are able, by tending toward different tubes, to turn different parts of the body all in the same direction. So in the present case [see fig. 63], spirits issuing from the same 183 point *b* tend towards tubes 4, 4, and 8, simultaneously turning the two eyes and the right arm toward object B.

You will readily accept this if, in order to understand what the idea of the distance of objects consists in, you assume that as the gland's position changes, the closer points on its surface are to the centre of the brain *o*, the more distant are the places corresponding to them, and that the further the points are from it the closer the corresponding places are. Here, for example, we assume that if *b* were pulled further back, it would correspond to a place more distant than B, and if it were made to lean further forward it would correspond to a place that was closer.

And when a soul has been put in this machine, this will allow it to sense various objects by means of the same organs, disposed in the same way, and without anything at all changing except the position of the gland H. Here [fig. 65], for example, the soul can sense what is at point L by means of the two hands holding sticks NL and OL, because it is from point L on gland H that the spirits entering tubes 7 and 8 issue. Now suppose that gland H leans a little further forward, in such a way that points *n* and *o* on its surface are at the places marked *i* and *k*, and that as a consequence it is from them that the spirits entering 7 and 8 issue: the soul would sense 184 what is at N and what is at O by means of the same hands without them being changed in any way.

Moreover, it should be noted that when gland H is inclined in one direction by the force of the spirits alone, without the aid of either the rational soul or the external senses, the ideas which are formed on its surface derive not only from inequalities in the tiny parts of the spirits causing corresponding differences in the humours, as mentioned earlier, but also from the imprints of memory. For if the figure of one object is imprinted much more distinctly than that of another at that place in the brain towards which this gland is properly inclined, the spirits issuing

Fig. 65

from it cannot fail to receive an impression of it. And it is in this way that
past things sometimes return to thought as if by chance and without the
memory of them being stimulated by any object impinging on the senses.

But if many different figures are traced in this same region of the brain
each almost as perfectly as the other, as commonly happens, the spirits
will receive something from the impression of each of them, and this will
occur to a greater or lesser degree depending on the various ways in
which the parts of the figures match one another.[51] It is in this way that

[51] The Epicureans had developed a detailed account of how our perceptions can become 'mixed' if
the atoms emitted from different bodies interfere with one another, and it is possible that Descartes
is following this traditional atomist account here.

Fig. 66

chimeras and hippogriffs are formed in the imagination of those who dream while awake, that is, those who let their fancy nonchalantly wander here and there without external objects diverting it, and without being directed by reason.

But the effect of memory that seems to me to be most worthy of consideration here is that, without there being any soul present in this machine, it can naturally be disposed to imitate all the movements that real men – or many other similar machines – will make when it is present.

The second cause that can determine the movements of gland H is the action of objects impinging on the senses. For it is easy to grasp that when the degree to which tubes 2, 4, and 6 [fig. 66] are open is increased by the action of the object ABC, the spirits, which immediately begin to flow toward them more freely and rapidly than they would otherwise, pull the gland with them a little, and cause it to lean, if nothing else prevents it from doing so; so that, changing the disposition of its pores, it begins to direct a much greater quantity of spirits through *a*, *b*, and *c* to 2, 4, and 6 than it did previously, and this makes the idea that these spirits form

185

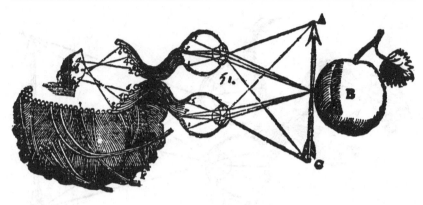

Fig. 67

correspondingly more perfect. This is the first effect that I want you to take note of.

The second is that, while this gland is leaning thus to one side, this hampers the ease with which it is able to receive ideas of objects acting on the other senses. In this case [see fig. 66], for example, during the time when almost all the spirits that gland H produces leave from points *a*, *b*, and *c*, there are not enough leaving point *d* to form there an idea of object D, whose action I assume to be neither as lively nor as strong as that of

186 object ABC. From this you see how ideas impede one another, and why we cannot be very attentive to several things at the same time.

It must also be noted that in the sense organs, when they are first impinged upon more strongly by one object than others, but are not yet as disposed as they might be to receive its action, the presence of the object is enough to succeed in disposing them completely toward it. Thus, for example, if the eye is disposed to look at a very distant place, when a very near object ABC begins to come into view, I maintain that the action of this object will be able to change the disposition of the eye immediately in such a way that it fixes upon the object.

And so that you might understand this more easily, consider first the difference between an eye disposed to look at a distant object as in [fig. 60] and the same eye disposed to look at a nearer object, as in [fig. 67]; this difference consists not just in the crystalline humour's being a little more arched and in the other parts of the eye being correspondingly differently disposed in the earlier figure than in the present one, but also in the tiny tubes 2, 4, and 6 being inclined toward a nearer point, in the gland H being a little more advanced towards these tiny tubes, and in the region *a, b, c* of the gland's surface being proportionately a little more arched or curved so that, in both figures, spirits issuing from point *a* always tend toward tube 2, with those from *b* tending toward tube 4 and those from *c* tending toward tube 6.

Consider also that the movements of gland H are sufficient in them- 187 selves to change the position of these tubes, and consequently that of the whole eye. As I have already said, the main point is that these tubes can make all the bodily parts move.

Next consider that the tubes 2, 4, and 6 [see fig. 67] can be opened by the action of object ABC in proportion to how much the eye is disposed to look at it. If the rays that fall on point 3, for example, all come from point B, as they do when the eye looks at it fixedly, it is evident that their actions must pull more strongly on fibre 3/4 than if some came from A, some from B, and some from C (as they would if the eye were slightly differently disposed, for in this case their actions, not being as similar or as unified, could not be anything like as strong, and can often even impede one another). This only occurs when the outlines of objects are neither too similar nor too indistinct, and also happens in the case of objects whose distance and parts the eye can distinguish, as I have pointed out in the *Dioptrics*.[52]

Moreover, consider that the gland H can be moved very much more easily when it is inclined in the direction in which the eye will be disposed to receive the action of whatever object is already acting most strongly upon it, than in the contrary direction. Thus, for example, in [fig. 60] where the eye is disposed so as to look at a distant object, less force will be required to make it incline itself slightly forward than backward, 188 because in inclining backward the eye would be less disposed than it was to receive the action of the object ABC, which we assume to be nearby and to be the object acting most strongly against it. And this would cause the tiny tubes 2, 4, and 6 to be opened less by the action of the object; and the spirits issuing from the points *a, b,* and *c* would flow much less freely toward these tubes. Whereas when it leant forward, the eye would be better disposed to receiving this action, the tiny tubes 2, 4, and 6 would open more, and consequently spirits issuing from points *a, b,* and *c* would flow toward them very much more freely, so that, as soon as the gland

---

[52] *Dioptrics*, Discourse 6.

begins to move to even the slightest degree, the flow of these spirits immediately bears it along, not allowing it to stop until it is fully disposed in the manner that you see in fig. 67, where the eye is looking fixedly at this nearby object ABC.

So that it remains only for me to tell you what cause can initiate its movement. In the normal course of things, this is just the force of the object itself, which, acting against the sense organ, makes certain tiny tubes in the inside surface of the brain open up more so that the spirits, which immediately begin to flow towards these tiny tubes, pull the gland with them and make it lean in that direction. But where these tubes have already been opened to the same or a greater extent than this object would have opened them, we must consider that the tiny parts of the spirits that flow through its pores, being unequal, push it strongly here and there in the blink of an eye, without a moment's respite. And if they should first happen to push it in a direction in which it can only incline with difficulty, their action, because it is not very strong in itself, can have hardly any effect at all. On the other hand, as soon as they push it slightly in the direction in which it is already being carried, it cannot help but be inclined in that direction immediately, and as a result dispose the sense organ to receive the action of its object in the most perfect way possible, as I have just explained.

Let us now leave the conduction of the spirits to the nerves, and look at what movements depend on them. If none of the tiny tubes on the inside surface of the brain is more open than any other, nor differs in another way, and as a consequence the spirits have no impression of any particular idea in them, they will spread out indifferently in all directions and pass from the pores in the vicinity of B [see fig. 58] toward those in the vicinity of C, whence their subtlest parts flow directly from the brain through the pores of the minute membrane which envelops it, while the remainder, making their way toward D, will proceed into the nerves and the muscles, without causing any particular effect there, because they will be distributed to all muscles equally.

But if there are some tubes that are opened to some degree, or just opened in a different way to their neighbours, through the action of the objects moving the senses, then the tiny fibres composing the substance of the brain, some of which will consequently be a little more tense or relaxed than others, will conduct the spirits toward regions at its base and from there to some nerves with more or less force than to others. And this

189

190

will be enough to cause different movements in the muscles, in accord with what has already been explained fully.

Now I want you to think of these movements as being similar to those to which we men are incited by the various actions of objects impinging on our senses, and to this end I want you to consider six different conditions to which the different movements may be due. The first is the place from which the action that opens those tiny tubes through which the spirits first enter proceeds. The second is the force, and all the other qualities of this action. The third is the disposition of the tiny fibres that make up the brain. The fourth is the unequal force that the different parts of the spirits can have. The fifth is the different positions of the external bodily parts. And the sixth is the interconnection between the many actions that move the senses at the same time.

As regards the place from which the action proceeds, you already know that if the object ABC [see fig. 67], for example, were to act on some sense other than vision, it would open tubes in the inside surface of the brain other than those marked 2, 4, and 6. And if it were closer to or farther ₁₉₁ away from the eye, or located elsewhere in respect to it, it could as a matter of fact open the same tubes, but they would have to be located elsewhere and therefore would be able to receive spirits from other points of the gland than those marked *a*, *b*, and *c*, and conduct them to regions other than ABC, where they conduct them now, and so on.

As regards the various qualities of the action that opens these tubes, you also know that because of differences in these qualities they open them in different ways; and we must consider this alone to be enough to change the course of the spirits in the brain. For example, if object ABC is red, that is, if it acts on the eye 1, 3, 5 in the way I said above was required to make it sense the colour red, and if in addition the object has the shape of an apple or some other fruit, we must consider that it will open the tubes 2, 4, and 6 in a particular way which will cause the parts of the brain near N to press against one another a little more than they usually do, with the result that spirits entering through tubes 2, 4, and 6 will make their way from N through *o* toward *p*. Whereas if the object ABC had a different colour or shape, it would not be the tiny fibres near N and *o* that would deflect the spirits entering 2, 4, and 6 but other, neighbouring ones.

And if the heat of the fire A [fig. 68], which is close to hand B, were only moderate, we would have to conclude that the way in which it would open tube 7 would cause the parts of the brain near N to press together, and ₁₉₂

Fig. 68

those near *o* to be spread apart a little more than usual, and because of this the spirits coming from tube 7 would go from N through *o* to *p*. But if we assume that this fire burns the hand, we must consider that its action opens tube 7 so wide that the spirits entering it are sufficiently strong to pass further, in a straight line, beyond N, namely as far as *o* and R where, pushing before them the parts of the brain they find in their way, they push in such a way that they encounter resistance and are deflected toward S, and so on.

As for the disposition of the tiny fibres that make up the substance of the brain, it is either acquired or natural; and since what is acquired depends upon all the various circumstances that change the course of the spirits, I shall be able to explain them better later. But in order to show you in what the natural ones consist, consider that, in forming them, God so disposed these tiny fibres that the passages He left between them are able to conduct the spirits, when these are moved by a particular action, toward nerves which allow in this machine just those movements that a similar action could incite in us when we follow our natural instincts. Thus, for example, if fire A [see fig. 68] burns hand B and causes the spirits entering tube 7 to tend toward *o*, the spirits find there two pores or principal passages *o*R, *os*. One of these, namely *o*R, conducts them into all the nerves that serve to move external bodily parts in the way needed to 193 avoid the force of this action, such as those that withdraw the hand or the arm or the entire body, and those that turn the head and the eyes toward the fire so as to see more particularly what it must do in order to protect itself. And through the other, *os*, they enter all those that serve to cause inner emotions, like those that pain occasions in us: these are nerves such as those that constrict the heart, agitate the liver, and so on. And they even also enter those nerves causing the external movements which bear witness to these, such as those that excite tears, or wrinkle the forehead and cheeks, or dispose the voice to cry. On the other hand, if hand B were very cold and fire A were to warm it moderately without burning it, this would cause the same spirits entering through tube 7 to proceed no longer to O and R, but toward *o* and *p*, where they would again find pores disposed in such a way as to conduct them into all the nerves which can serve for movements suited to this action.

And note that I have explicitly distinguished between the two pores *o*R and *os* in order to alert you to the fact that two kinds of movement almost always follow every action: namely, external movements that serve either in the pursuit of desirable things or in the avoidance of injurious ones, and internal movements that are commonly termed passions, which serve to dispose the heart, the liver, and all the other organs on which the temperament of the blood – and as a result, that of the spirits – depends, so that the spirits produced at a particular time are those suited to causing the external movements that must follow. For assuming that the 194 various qualities of these spirits are one of the circumstances that serve to change their course, as I shall explain in a moment, we may readily appreciate that if, for example, it is a question of avoiding some evil by force, by surmounting it or chasing it away, as the passion of anger inclines us to do, then the spirits must be more unevenly agitated and stronger than usual. And on the other hand, when one must avoid it by hiding or bearing it with patience, as the passion of fear inclines us to do, it must be less abundant and weaker. To achieve this, the heart must be constricted, and must spare and save the blood for when it needs it. And you can judge the other passions proportionately.

As for other external movements which serve neither to avoid evil nor to pursue the good but merely bear witness to the passions, such as those consisting in laughing or crying, these occur only by chance because the nerves through which the spirits enter in order to produce them originate very close to those through which spirits enter to give rise to the passions, as anatomy will show you.

But I have not yet shown you how the various qualities of the spirits can have the force to change the direction of their flow. This occurs chiefly when they are directed by others either only very slightly or not at all. For example, if the nerves of the stomach are agitated in the way I said earlier 195 they must be if they are to cause the sensation of hunger, and if nevertheless nothing is presented to any of the senses or to memory which appears to be edible, then the spirits that will be caused by this action to enter tube 8 in the brain will proceed to a region where there are many pores so disposed as to conduct them indifferently into all the nerves that can serve for the search or pursuit of some object, so that only the inequality of their parts can cause them to make their way through some rather than others.

And if it turns out that the strongest of these parts are those which now tend to flow toward certain nerves, and then immediately after towards their opposites, the machine will be imitating the movements seen in ourselves when we hesitate and are in doubt about something.

Similarly, if the action of the fire A lies somewhere between actions that can conduct the spirits toward R and those that can conduct them toward p, that is, between those causing pain and those causing pleasure, it is easy to understand that the inequalities between them are alone sufficient to direct the one or the other: just as the same action is agreeable to us when we are in a good humour and can displease us when we are sad and sorrowful. And from this you can deduce the basis for everything I said earlier about the humours or inclinations, whether natural or acquired, that depend on differences in the spirits.

196   As for the various positions of external parts of the body, one need only

consider that they alter the pores that carry these spirits immediately into the nerves. For example, if fire A burns hand B and the head is turned toward the left – instead of the right as it is at present – the spirits will still go, in the same way as they do now, from 7 to N, and then to *o*, and from there to R and *s*. But from R, instead of going to *x* – through which I am assuming they must pass if they are to hold the head upright when it is turned, as it is now, to the right – they will go to *z* – which I am assuming they would have to enter if they are to hold the head upright if it were turned to the left – especially as the present position of this head, which causes the tiny fibres of the substance of the brain near *x* to be more relaxed and easier to separate than those near *z*, when it is changed will make those at *z* very relaxed and those at *x* tense and tight.

Thus to grasp how a single action can, without changing, move now one foot of this machine, now the other, depending on which is required for it to walk, it is enough to consider that the spirits pass through a 197 single pore, the end of which is differently disposed, and so conducts them into different nerves, when the left foot goes forward than when the right one does. And this is applicable to everything I said earlier about respiration and similar movements which do not usually depend on an idea; and I say 'usually' because they may sometimes depend on them.

Now I believe I have given enough explanation of the functions of the waking state, and there remains only a few things to be said about sleep. Just cast your eyes on fig. 69 and see how the small fibres D, D that enter the nerves are relaxed and pressed together, and you will understand how, when this machine corresponds to the body of a sleeping person, the actions of external objects are for the most part prevented from reaching the brain and being sensed; and the spirits in the brain are kept from reaching the external bodily parts so as to move them. These are the two principal features of sleep.

As for dreams,[53] these depend in part on the unequal force that the spirits can have in issuing from gland H, and in part from the impressions that are involved in memory, so that the only way they differ from the 198 ideas that I said above are occasionally formed in the imagination of those who are awake is that the images formed in dreams can be more distinct and more lively than those formed during waking. The reason for this is

---

[53] As Hall points out in the notes to his translation of *The Treatise on Man* (82 n.131), 'Descartes appears to wish to link waking with general intracerebral tension or turgor, dreamless sleep with general laxness, and dreaming with local tension or turgor during a period of general laxness.'

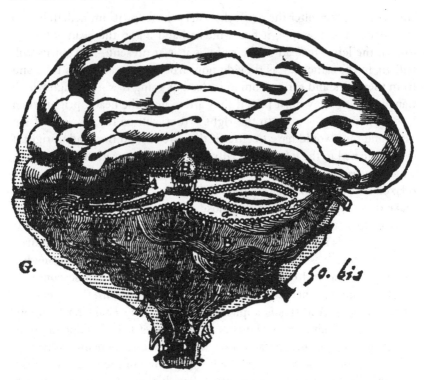

Fig. 69

that the same force will open the tiny tubes (such as 2, 4, and 6) and the pores (such as *a*, *b*, and *c*) which serve to form the images in question more widely when the surrounding parts of the brain are relaxed and loosened, as we can see in [fig. 70], than it does when they are tense, as you can see in earlier figures. And this also shows that if it happens that the action of some object impinging on the senses can pass as far as the brain during sleep it will not form the same idea that it would while it was awake, but will form there some other more noticeable and sensible one: as sometimes, while we sleep, if we are stung by a fly, we dream that we have been stabbed with a sword. Or if we are not properly covered, we imagine ourselves to be completely naked; and if we are covered a little too much, we think we are weighed down by a mountain.

Moreover, during sleep the substance of the brain, which is at rest, has the opportunity to nourish and repair itself, being moistened by the blood contained in the little veins or arteries that appear on its outside surface.

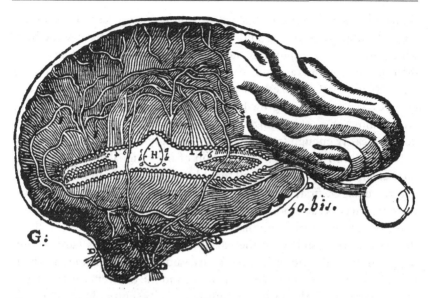

Fig. 70

So that, after some time, its pores having become narrower, the spirits 199 need less strength than previously to keep the substance of the brain tense: just as the wind does not need to be as strong to inflate a ship's sails when they are damp rather than dry. And yet these spirits are stronger inasmuch as the blood which produces them is purified in passing several times through the heart, as I have noted above. From this it follows that this machine must naturally wake itself up after it has slept for some time. And, reciprocally, it must also go to sleep again after it has been awake for some time, because during waking the substance of its brain is dried out, its pores being gradually enlarged by the continual action of the spirits, and in the meantime, if it happens to eat (which it will certainly do from time to time, if it can find something to eat, since hunger will excite it to do this), the juice of the food when mixed with the blood will make it coarser, and consequently it will produce less spirits.

I shall not pause to tell you how noise and heat, and other actions which very forcefully move the internal parts of the brain through the media-tion of the sense organs, or how joy and anger and the other passions that greatly agitate the spirits, or how the dryness of the air, which renders the blood more subtle, or similar circumstances, can prevent it sleeping: nor on the other hand how silence, sadness, the humidity of the air and 200 similar things can invite it to sleep. Nor how a great loss of blood, too much fasting or drinking, and other excesses which have something which increases or diminishes the strength of the spirits, depending on their different temperaments, make the machine either wake or sleep too much. Nor how through excessive waking its brain can be weakened, and by an excess of sleeping grow heavy like one who is senseless or stupid. Nor innumerable other things: since it seems to me that they can all be deduced easily enough from what I have already explained.

Now before I pass to the description of the rational soul, I want you once again to reflect a little on all that I have just said about this machine; and to consider, first, that I have postulated in it only such organs and working parts as can readily persuade you that they are the same as those in us, as well as in various animals lacking reason. For in the case of those that are clearly visible with the naked eye, the anatomists have already observed them all, and as for what I have said about the manner in which the arteries carry the spirits into the head, and the difference between the inside surface of the brain and the substance at its centre, they will be able to see enough indications there to allay any doubts, if only they look a little more closely. They will no longer have any doubts about the tiny doors or valves that I placed in the nerves at the entrances to each muscle, if they notice that nature generally has formed these valves at those places in our body where matter regularly enters and could have a tendency to escape, as at the entrances to the heart, the gall-bladder, the throat, the large intestines, and the principal divisions of the veins. And as regards the brain, they will not be able to imagine anything more likely than my own view that it is composed of various tiny fibres connected together in different ways, in view of the fact that all skin and flesh seem similarly composed of many fibres or threads, and that one observes the same thing in plants, so that this is a property that seems to be common to all bodies which are able to grow and be nourished by the union and joining together of the tiny parts of other bodies. Finally, as for the rest of what I have supposed with regard to things not perceivable by any sense, they are all so simple and common, and even so small in number, that if you compare them with the diversities of composition and marvellous ingenuity evident in the structure of the visible organs, you will be more inclined to think that I have omitted many that are in us, rather than having included some that are not. And knowing that Nature always acts by the simplest and easiest means, you will perhaps conclude that it is not possible to find anything more like those she uses than the ones proposed here.

Further, I desire that you consider that all the functions that I have attributed to this machine, such as the digestion of food, the beating of the heart and the arteries, the nourishment and growth of the bodily parts, respiration, waking and sleeping; the reception of light, sounds, odours, smells, heat, and other such qualities by the external sense organs; the impression of the ideas of them in the organ of common sense and the imagination, the retention or imprint of these ideas in the memory; the internal movements of the appetites and the passions; and finally the external movements of all the bodily parts that so aptly follow both the actions of objects presented to the senses, and the passions and impressions that are encountered in memory: and in this they imitate as perfectly as is possible the movements of real men. I desire, I say, that you should consider that these functions follow in this machine simply from the disposition of the organs as wholly naturally as the movements of a clock or other automaton follow from the disposition of its counterweights and wheels. To explain these functions, then, it is not necessary to conceive of any vegetative or sensitive soul, or any other principle of movement or life, other than its blood and its spirits which are agitated by the heat of the fire that burns continuously in its heart, and which is of the same nature as those fires that occur in inanimate bodies.

The Description of the Human Body and All Its Functions, those that do not depend on the Soul as well as those that do. And also the principal cause of the formation of its parts.

[Part 1. Preface]

There is no more fruitful occupation than to try to know oneself. And the benefit that one expects from this knowledge does not just extend to morals, as many may initially suppose, but also to medicine in particular. I believe one can find very many reliable precepts in medicine, as much 224 for curing illness as for preventing it, and even also to slow the course of ageing, so long as one has studied sufficiently to know the nature of our body, not attributing to the soul functions which depend only on the body and on the disposition of its organs.

But because it is the experience of everyone from childhood that many of our movements obey the will, which is one of the powers of the soul, this has disposed us to believe that the soul is the principle behind all of them. And the ignorance of anatomy and mechanics has contributed to this, for in considering only the exterior of the human body, we never imagined that it had enough organs or springs in it to move itself in all the different ways in which we see it move. And we have been confirmed in this error in judging that dead bodies have the same organs as living ones, for they lack nothing but the soul and yet there is no movement in them.

When we make the attempt to understand our nature more distinctly, however, we can see that our soul, in so far as it is a substance distinct from body, is known to us solely from the fact that it thinks, that is to say, understands, wills, imagines, remembers, and senses, because all these functions are kinds of thoughts. Also, since the other functions that are attributed to it, such as the movement of the heart and the arteries, the digestion of food in the stomach, and such like, which contain in them- 225 selves no thought, are only corporeal movements, and since it is more common for one body to be moved by another body rather than by the soul, we have less reason to attribute them to the soul than to the body.

We can also see that when parts of our body are harmed – when a nerve is pricked, for example – the upshot of this is that not only do they stop obeying our will (which is what they normally do) but often even have convulsive movements, which are quite opposed to it. This shows that the soul can cause no movement in the body unless all the corporeal organs required for that movement are properly disposed. And when the body has all the organs disposed for this movement, it does not need the soul to produce it. Consequently, all those movements that we do not experience as depending on our thought must not be attributed to the soul but only to the disposition of our organs; and even those movements that are called 'voluntary' proceed principally from this disposition of the organs, for they cannot have been produced without it, no matter how much we will it, and even though it is the soul that determines them.

Furthermore, although all these movements cease in the body when it dies and the soul leaves it, it should not be inferred from this that it is the soul that produces them, but only that the body's no longer being able to produce them and the soul's leaving it are due to the same cause.

It is true that it may be hard to believe that the disposition of organs 226 alone is sufficient for the production in us of all the movements that are not determined by our thought. This is why I shall try to demonstrate this here, and to explain the entire machine of our body in such a way that we will have no more reason to think that it is our soul that excites in us those movements that we do not experience as being directed by our will, than we have to judge that there is a soul in a clock that makes it tell the time.

There is no one who does not already have some knowledge of the different parts of the body, that is, who does not know that it is composed of a great number of bones, muscles, nerves, veins, arteries, and that it has a heart, a brain, a liver, lungs, and a stomach. And everyone has, at one time or another, seen various animals opened up, and gazed on the shape and arrangement of their interior parts, which are very much like our own. One need have studied no more anatomy than this to understand this book, for I shall explain any further details as the need arises.

First, I want the reader to have a general conception of all of the machine that I shall be describing. I shall say here that the heat that it has in its heart is like the great spring or the principle of all its movements, and that the veins are the tubes which conduct the blood from all the parts
227 of the body towards the heart, where it fuels the heat there; just as the stomach and the intestines are another much larger tube, perforated with many little holes, through which the juices from the food run through the veins, which carry them straight to the heart. And the arteries are yet another set of tubes, through which the blood, heated and rarefied in the heart, passes from there into all the other parts of the body, to which it brings heat and matter to sustain them. Finally, the most agitated and most active parts of this blood are carried to the brain by arteries which follow the straightest line in their passage from the heart, comprising an air or very fine wind which is called the 'animal spirits'. These dilate the brain, enabling it to receive both the impressions from external objects, and those from the soul, thereby acting as the organ or the seat of the common sense, of the imagination, and of the memory. Then, this same air or these same spirits flow from the brain through the nerves into all the muscles, thereby making these nerves serve as organs of the external senses, and inflate the muscles in various ways imparting movement to all bodily parts.

In sum, these are all the things that I shall describe so that, knowing distinctly what in each of our actions depends only on the body, and what depends on the soul, we can make better use of the one and the other to heal or prevent their maladies.

228   *[Part 2. On the motion of the heart and the blood]*

It is beyond doubt that there is heat in the heart, for one can even feel it with one's hand when one opens up the body of a living animal. And we should not imagine that this heat is of a different nature from that which is caused by the addition of some fluid, or yeast, which causes the body with which it is mixed to expand.

But because the dilation of the blood that causes this heat is the first and principal spring of our whole machine, I would like those who have never studied anatomy to take the trouble to look at the heart of some land animal, something reasonably large (for they are more or less similar to those of men), and, having first cut off the end of the heart, to take note that there are two caverns or cavities inside, which are able to hold a lot of blood. If one then puts one's fingers in these cavities, towards the base of the heart (and from which it discharges its contents), to seek out the openings through which they receive the blood, what one will find there is that there are two very large ones in each: to wit, in the right ventricle, there is a large opening which leads the finger into the vena cava, and another which will lead it into the pulmonary artery. Then, if they cut through the flesh of the heart along this ventricle, as far as these two 229 openings, they will find three small membranes (commonly called the 'valvules') at the entry to the vena cava, which are arranged in such a way that when the heart is elongated and deflated (as it always is when animals are dead) they do not stop any of the blood from this vein descending into this ventricle; but if, because of the abundance and expansion of the blood that it contains, the heart is swollen and shortened, these three membranes must raise themselves and in this way close the entrance of the vena cava so that the blood can no longer descend through it into the heart.

Three small membranes or valvules can also be found at the entrance to the pulmonary artery, and these are differently disposed than those of the vena cava, so that they prevent the blood contained in this pulmonary artery from being able to descend into the heart; but if there is some blood in the right ventricle of the heart that tries to leave it, they will not prevent this leaving at all.

In the same way, if one puts one's finger into the left ventricle, one will find there two openings towards the base, which lead, one into the pulmonary vein, the other into the aorta. And in opening up this whole ventricle, we see two valvules at the entrance of the pulmonary vein which are just the same as those in the vena cava, and are positioned in the same way, and there would be no difference at all, were it not that the pulmonary vein is pressed on the one side by the aorta and on the other by the pulmonary artery, which makes its opening oblong. Because of this, two small membranes are enough to close it, rather than the three needed 230 to shut the vena cava.

One will also see three other valvules at the entrance to the aorta, which do not differ at all from those at the entrance to the pulmonary artery, so that they do not prevent the blood in the left ventricle of the heart rising into this aorta, but they do prevent it passing back down this artery into the heart.

And it will be noted that these two vessels, namely the pulmonary artery and the aorta, are composed of skin that is much stronger and thicker than the vena cava and the pulmonary vein. This shows that the latter have a completely different use from the former, and that what is called the 'venous artery' is really a vein [viz. the pulmonary vein], just as what is called the 'arterial vein' is really an artery [viz. the pulmonary artery]. But what made the ancient writers call an 'artery' what they should have called a 'vein', and call a 'vein' what they should have called an 'artery', is the fact that they believed that all the veins came from the right ventricle of the heart, and all the arteries from the left.

Finally, it will be noted that these two parts of the heart which are called its 'auricles' are nothing but the extremities of the vena cava and the pulmonary vein, which are widened and folded up here for reasons I shall go into below.

When the anatomy of the heart is seen in this way, if one considers that it always has more heat in it when the animal is alive than any other part of the body, and that the blood is of such a nature that when it is a little hotter than usual it expands very quickly, one cannot doubt that the movement of the heart, and following it the pulse, or the beating of the arteries, occurs in the way that I shall describe.

When the heart is elongated and deflated, there is no blood in its ventricles, except for a small amount which remains from that which has previously been rarefied. This is why two large drops enter them there, one falling from the vena cava into its right ventricle, and the other falling from the pulmonary vein into the left one, and the small amount of rarefied blood that remains in these ventricles, mixing straightaway with the fresh blood coming in, is like a kind of yeast, which causes it to heat and expand immediately, and by these means the heart swells, hardens, and becomes a little squatter in shape; and the little membranes at the entrances to the vena cava and the pulmonary vein rise and shut them in such a way that the blood is no longer able to descend from these two veins into the heart, and the blood that expands in the heart cannot rise towards these two veins. But it rises easily from the right ventricle into the pulmonary artery, and from the left into the aorta, without the small membranes at their entrances acting to prevent this.

And because this rarefied blood requires much more room than there is in the ventricles of the heart, it enters into the two arteries with great force, and by these means swells and rises at the same time as the heart; and it is this movement, as much of the heart as of the arteries, that is called the pulse.

Immediately after the blood, thus rarefied, has taken its course into the arteries, the heart deflates, becoming flabby and elongated, which is why so little blood remains in its ventricles; and the arteries deflate also, in part because the outside air, which is much closer to their branches than it is to the heart, makes the blood that they contain cooler and condenses it, and in part because there is about as much blood continually leaving them as there is entering them. Although it seems that, when the blood no longer rises from the heart into the arteries, their contents must go back down into the heart, in fact it cannot enter its ventricles, because the small membranes at the entrances to their arteries prevent it from doing so. It enters it rather from the vena cava and the pulmonary vein which, expanding in the same way as before, makes the heart and the arteries 233 move a second time, and thus their beating always continues while the animal is alive.

As for those parts that are called the 'auricles of the heart', their movement is different from that of the heart itself, but follows it very closely, for as soon as the heart is deflated, two large drops of blood fall into its ventricles, one from its right auricle, which is the extremity of the vena cava, the other from its left auricle, which is the extremity of the pulmonary vein, and by these means the auricles deflate. And the movement of the heart and the arteries, which then immediately inflate, to some extent inhibits the blood which is in the branches of the vena cava and the [pulmonary vein] from coming to fill these auricles, in such a way that they deflate; and instead of the heart inflating all at once, and then deflating gradually, the auricles deflate more rapidly than they inflate. Moreover, the movement by which they inflate and deflate is confined to them, and does not extend to the vena cava and the pulmonary vein of which they are the extremities, and this is why they are so much larger, and otherwise bent, and made up of much thicker and fleshier membranes than these other two veins.

But in order that all this be understood better, we must consider more particularly the material of the four vessels of the heart. And first, as regards the vena cava, we should note that it extends throughout all parts of the body except the lung, so that all the other veins are only its 234 branches; for even the portal vein, which is spread throughout the spleen and the intestines, is joined to it so clearly by tubes in the liver that it can be included. One must thus consider all these veins as a single vessel which is named the vena cava at the spot where it is largest, and which always contains the major part of the blood that is in the body, which it naturally conducts into the heart, so that if it were to contain only three drops, these would leave the other parts and would proceed towards the right auricle of the heart. The reason for this is that the vena cava is much larger here than anywhere else, and it goes from there by narrowing gradually as far as the ends of its branches; and the membranes from which its branches are composed can be stretched more or less according to the quantity of blood that they contain, always contracting some small part of itself by which means it drives this blood towards the heart. And, finally, there are valvules in several parts of its branches, which are so arranged that they completely close the passage, preventing the blood from flowing to their extremities, and thus becoming too distant from the heart when it comes about that its weight or some other cause pushes it there; but they do not prevent it flowing from the extremities towards the heart. Because of this, we must also judge that their fibres are also so arranged that they allow the blood to flow more easily in this direction than in the contrary one.

235   As regards the pulmonary artery and the pulmonary vein, we should note that these are also two vessels that are very large at the point at which they are attached to the heart, but that they divide very close to there into several branches, and these divide yet again into others which are very small. And they proceed by narrowing in proportion to their distance from the heart; each branch of one of these two vessels always accompanying some branch of the other, and also some branch of a third vessel, whose entrance is called the windpipe or the throat, and the branches of these three vessels do not go anywhere except the lung, which is made up of these alone, and they are so mixed together that one cannot point to any part of its flesh, which is large enough to be seen, in which each of these three vessels has none of its branches.

It should also be noted that these three vessels are different in that that whose entrance is in the throat never contains anything but respiratory air, and is made up of tiny cartilage and membranes very much harder than those that make up the other two. Similarly with the pulmonary artery, which is composed of membranes that are notably harder and thicker than those of the pulmonary vein, which are soft and slender just like those in the vena cava. This shows that, although these two vessels contain only blood, there is nevertheless a difference between them, in that the blood in the pulmonary vein is not as agitated, or driven with as much force, as that in the pulmonary artery. For, just as one sees that the hands of artisans become hard due to the manner of their use, so the cause of the hardness of the membranes and cartilage of which the windpipe is comprised is the force and agitation of the air that passes through it when one breathes. And if the blood were not more agitated when it enters the pulmonary artery than when it enters the pulmonary vein, the membranes of the former would be no thicker and harder than those of the latter.

But I have already explained how the blood enters the pulmonary artery with a force that is in proportion to how much it has been heated and rarefied in the right ventricle of the heart. It remains here only to say that, when this blood is dispersed through all the tiny branches of this pulmonary artery, it is cooled and condensed by the respiratory air; and because of this tiny branches of the vessel that contain this air are mixed among them in all parts of the lung; and the new blood that comes from the right ventricle of the heart in this same pulmonary artery enters it with such force that it drives that which has begun to condense and makes it pass at the extremities of its branches into the branches of the pulmonary vein, where it flows very easily towards the left ventricle of the heart.

And the main use of the lung consists in one thing alone: by means of the respiratory air, it thickens and tempers the blood that comes from the right ventricle of the heart before it enters the left ventricle; without this it would be too rare and too fine to serve to fuel the fire that it encounters there. Its other use is to contain the air that serves to produce the voice. Also, we see that fish and other animals that have only a single ventricle in the heart, all lack a lung, as a result of which, they are mute, so that none of them can make a sound. But they are also all of a very much colder constitution than animals that have two ventricles in their hearts, because the blood of these latter, having already been heated and rarefied once in the right ventricle, falls back into the left ventricle a little later where it stirs up a fire that is more lively and warmer than it would be were it to come immediately from the vena cava. And although this blood re-cools and condenses in the lung, nevertheless, because it remains there for a short time, and because it does not mix with any grosser matter there, it retains an ability to dilate and reheat better than that which it had before it entered the heart. Similarly, experience shows that oils that have been made to pass several times through a distillation flask are much easier to distil the second time than the first.

And the shape of the heart serves to demonstrate that the blood heats up more and expands with greater force in the left ventricle than in the right one; for it can be seen that it is very much larger and rounder, that the flesh surrounding it is thicker, and yet it is the same blood passing through this ventricle as passes through the other, and which is thinned because it has provided nourishment to the lung.

The openings of the vessels of the heart also serve to show that respiration is necessary for the condensation of the blood in the lung; for 238 it can be seen that infants, who cannot breathe while they are in their mother's womb, have two openings in the heart which are not to be found in those which are older. Through one of these openings, the blood from the vena cava runs with that from the pulmonary vein in the left ventricle of the heart; through the other (which is shaped like a small tube) a part of the blood that comes from the right ventricle passes from the pulmonary artery into the aorta, without entering the lung. One can also see that these two openings gradually close up by themselves when the infant is born and is able to breathe; by contrast, in geese, ducks, and similar animals, which can remain a long time under the water without breathing, they never close up.

It remains to note, with respect to the aorta, that it is the fourth vessel of the heart, that all of the body's other arteries are not as large as it is, and are only its branches, through which the blood that it receives from the heart is promptly carried to all its limbs. And all these branches of the aorta are joined to those of the vena cava, just as those of the pulmonary artery are joined to the branches of the pulmonary vein; so that, after having distributed to all parts of the body what they need from the blood, whether it be for their nourishment or for other uses, they carry all the surplus in the extremities of the vena cava, where it once more runs towards the heart.

And thus the same blood goes backwards and forwards several times, 239 from the vena cava into the right ventricle of the heart, then from there via the pulmonary artery into the pulmonary vein, and from the pulmonary vein into the left ventricle, and from there via the aorta into the vena cava, this making a perpetual circular motion which would be enough to sustain the life of animals, without their needing to drink or eat, if none of the parts of the blood left the arteries or veins while it flowed in this fashion. But many parts continually leave it, and these are supplied by the juice of foods, which come from the stomach and intestines, as I shall explain below.

Now this circulatory movement of the blood was first observed by an English physician called Harvey, whom one cannot praise too highly for such a useful discovery. And although the ends of the veins and the arteries are so delicate that one cannot see with the naked eye the openings by which the blood passes from the arteries into the veins, there are nevertheless several places where it can be seen: above all in the great vessel which is made up of layers of the larger of the two membranes that envelop the brain, in which many veins and many arteries are found, so that the blood is led there through the latter, then returning through the former to the heart. This can also be seen to some extent in the spermatic veins and arteries. And the evidence showing that the blood passes in this way from the arteries into the veins is so strong that they leave one no room for doubt.

For if, having opened the chest of a living animal, one ties the aorta sufficiently close to the heart, so that no blood can descend from its branches, and if one cuts between the heart and the tie, all the blood of 240 this animal, or at least the greater part of it, quickly escapes via this opening. This would be impossible if that in the branches of the aorta did not have passages by which to enter into the branches of the vena cava, from where it passes into the right ventricle, and from there into the pulmonary artery, at the extremities of which it must also find passages in order to enter the pulmonary vein, which leads it into the left ventricle, and from there into the aorta, from where it leaves.

If one does not wish to take the trouble to open up a living animal, one need only consider the way in which surgeons usually tie the arm to bleed it, for if they tie it quite tightly a little higher, that is to say a little closer to the heart than the point at which they open the vein, the blood gushes out in much greater quantities than if it had not been tied. But if it is tied too tightly, the flow is stopped, just as it is if they tie it a little further from the heart – but not at the place where the vein opens – even if they do not tie it very tightly.

This clearly shows that, in its ordinary course, the blood is carried towards the hands and other extremities of the body by the arteries, and returns from these through the veins towards the heart. And this has already been so clearly demonstrated by Harvey, that it can be doubted only by those so attached to their prejudices, or so accustomed to dispute everything, that they cannot distinguish true and certain arguments from those that are false or probable.

But I believe Harvey has not been so successful as regards the movement of the heart, for he thinks, contrary to the common opinion of other physicians, and against the common judgement of sight, that when the heart lengthens, its ventricles increase in size, and conversely that when they shorten, they become narrower. Against this, I shall demonstrate that they always become larger.

The arguments that have led him to this view are as follows. He has observed that the heart, in shrinking, becomes harder, and also that in frogs and other animals that have little blood, it becomes more white, or less red, than when it lengthens; and that, if one makes an incision down as far as the ventricles, it is at the moment when it is shrunk that the blood leaves through the incision, and not when it is elongated. From this he believes we must conclude that, since the heart becomes hard, it is contracting; and moreover that its becoming less red in some animals indicates that the blood is leaving it. Finally, he thinks that since we see this blood leave via the incision, we must consider that the blood comes when the place that contains it is narrower.

He could have confirmed this by a very evident experiment: namely, if one cuts the point of the heart of a living dog, and through the incision one puts one's finger into one of its ventricles, one will clearly feel that every time the heart shortens it presses the finger, and that it will stop pressing it whenever it is elongated. This seems to ensure conclusively that its ventricles are narrower when the finger is pressed more than when it is pressed less. But all this shows that the same observations can often mislead us, if we do not examine their possible causes sufficiently. For although, if the heart does contract from within, as Harvey believes, this would make it become harder and less red in animals that have little blood, and would also make the blood in the ventricles spurt out through the incision we have made, and, finally, would make the finger inserted in the incision feel pressure; nevertheless, none of this alters the fact that the same effects could also proceed from a different cause, namely the expansion of the blood as I have described it.

But in order to be able to tell which of these two causes is the true one, we must consider other observations which are not compatible with both of them. And the first that I can give is that if the heart hardens due to a contraction of the fibres in it this would necessarily reduce its size; but if the hardening is due to the expansion of the blood contained in the heart, this on the contrary would lead to an expansion. Now observations show us that it loses nothing of its size: rather, it grows larger, which has led other physicians to consider that it swells up during this phase. It is true nevertheless that the increase in size is not great, but the reason for this is clear: the heart has several fibres stretched like cords from one side of its ventricles to the other, and these prevent them from opening very much.

Another observation shows that when the heart shortens and hardens, 243 its ventricles do not become narrower but, on the contrary, become larger: if one cuts the point of the heart of a young rabbit that is still alive, the naked eye shows that its ventricles become a little larger and expel blood at the moment at which the heart hardens, and even when it expels only small drops of blood, because very little blood remains in the animal's body, they continue to have the same size. And what prevents them from opening ever wider are fibres which stretch from one side to another, which hold them in place. What makes this much less apparent in the heart of a dog or some other more vigorous animal than in a young rabbit is that the fibres take up more of the ventricles; they stiffen when the heart hardens and can press against a finger inserted into one of the ventricles. But despite that, the ventricles do not become narrower but on the contrary larger.

I would add yet a third observation, which is that the blood does not leave the heart with the same qualities it had when it entered it, but is very much warmer, more rarefied, and more agitated. Now supposing that the heart moves in the way that Harvey describes, not only must we imagine some faculty which causes the movement, the nature of which is much more difficult to conceive than what it is invoked to explain: we must also suppose the existence of yet other faculties that alter the qualities of the 244 blood while it is in the heart. But if we confine our attention instead to the expansion of the blood, which must necessarily follow its heating, which everyone recognises is greater in the heart than in any other part of the body, then it will be clear that this expansion is enough to make the heart move in the way I have described, and also to change the nature of the blood in the way observation indicates. Indeed, it is also sufficient to explain any change one might imagine as necessary so that the blood is prepared and made more suitable for nourishing all the bodily parts that can be used for all the other functions for which it is used in the body. In this way, we need suppose no unknown or extraneous faculties.

For what better and swifter arrangement can we imagine than that which is brought about by fire, which is the most powerful agent we know in nature: rarefying the blood, it separates its small parts from one another, dividing them up and changing their shapes in every imaginable way.

This is why I am extremely surprised that, despite the fact that it has always been known that there is more heat in the heart than in the rest of the body, and that the blood can be rarefied by heat, it has not been noticed by anyone to date that it is this rarefaction of the blood alone that is the cause of the movement of the heart. For although it might seem
245 that Aristotle thought this when he wrote in chapter 20 of his book *De respiratione* 'that this movement resembles the action of a liquid that heat brings to a boil', and also that what causes the pulse is 'juices from the food one has eaten continually coming into the heart and rising to its outer wall', nevertheless, because he makes no mention in this passage of the blood, or of the material from which the heart is constructed, it is clear that it is just by chance that he has said something approaching the truth, and that he possessed no certain knowledge of it. Nor was his opinion adopted by anyone, even though he had the good fortune to have a number of followers on many other questions where his views are far less plausible.

Yet it is so important to know the true cause of the heart's movement that, without it, we cannot know anything about the theory of medicine, because all the other functions in the animal depend on it, as will be seen clearly from what follows.

[Part 3. On nutrition]

When one knows that the blood is continually dilated in the heart in this way, and from there pushed forcibly through the arteries into all the other
246 parts of the body, whence it returns subsequently through the veins towards the heart, one can easily judge that it is while it is in the arteries, rather than while it is in the veins, that it serves to nourish the parts of the

body. For although I do not wish to deny that, while it is flowing from the extremities of the veins to the heart, some of its parts pass through the pores in their surrounding membranes and become attached there – as happens particularly in the liver, which is without doubt nourished by the blood from the veins, because it receives almost none from the arteries – nevertheless in all other cases where the veins are accompanied by arteries, it is evident that the blood contained in these arteries, being finer and moved with a greater force than that in the veins, is very much more easily attached to other parts, without the thickness of their covering membranes hampering it. This is because, at their extremities their skins are hardly any thicker than those of the veins, and also because when the blood coming from the heart inflates them, the pores in their skins are enlarged in the process. And then the small parts of this blood, which have been separated from each other by the rarefaction that it has undergone in the heart, push these membranes forcefully from all sides, easily entering those pores of similar proportions, and go to strike the roots of the small filaments that make up the solid parts. Then, at the moment when the arteries deflate, these pores contract, and in the process several parts of the blood remain caught against the roots of the small filaments of the solid parts that they are nourishing (and several others flow away through the pores that surround them), and in this way they also enter into the 247 composition of the body.

But in order to understand this more distinctly, we must bear in mind that the parts of those living bodies that are maintained through nourishment, that is, animals and plants, undergo continual change, in such a way that the only difference between those that are called 'fluids', such as the blood, humours, and spirits, and those that are called 'solids', such as bone, flesh, nerves, and membranes, is that the latter move much more slowly than the others.

And in order to understand how these corpuscles move, we must think of all the solid parts being made up exclusively of small filaments which stretch out and fold back, and which are sometimes also intertwined, each emerging from somewhere on one of the branches of an artery. And we must think of the fluid parts, that is to say the humours and the spirits, flowing along these filaments, through the spaces that are found around them, making up infinitely many small channels which have their source in the arteries, and usually flowing from the pores of those arteries closest to the root of the filaments along which they run. And after following these filaments and various twists and turns in the body, they come finally to the surface of the skin, through the pores of which these humours and spirits evaporate into the air.

Now as well as these pores through which the humours and the spirits run, there are also many other narrower pores through which there 248 continually passes matter of the first two elements, as described in my *Principles of Philosophy*. And as the agitated matter of the first two elements encounters that of these humours and spirits, running along the filaments that make up the solid parts, they continually make the filaments move forward slightly, albeit very slowly; so that as a result every part of the filaments runs from where it has its roots to the surface of the limb where they terminate, and when it reaches there it comes into contact with the air or other bodies touching the surface of the skin, and separates from it. Thus there is always some part being separated from the end of each filament while at the same time another part is being attached to the root, in the manner already described. But the separated part evaporates into the air if it emerges from the skin, whereas if it emerges from the surface of a muscle, or from some other internal part, it mixes with the fluid parts and flows with them wherever they go: sometimes outside the body, and sometimes through the veins towards the heart, to which the fluid parts often return.

Hence it can be seen that all the parts of the filaments making up the solid parts of the body undergo a motion which is no different from that of the humours and spirits, only slower; similarly, the motion of the humours and spirits is slower than that of the most subtle matter.

249 And these differences in speed cause these various solid or fluid parts, in rubbing against one another, to become smaller or larger, and to behave in different ways depending on the particular constitution of each body. When one is young, for example, because the filaments that make up the solid parts are not joined to one another very firmly, and the channels along which they flow are quite large, the motion of these filaments is not as slow as when one is old, and more matter is attached to their roots than is detached from their extremities, which results in their becoming longer and stronger, and their increase in size is the means by which the body grows.

When the humours between these filaments do not flow in great quantity, they all pass quite quickly along the channels containing them, causing the body to grow taller without filling out. But when these humours are very abundant, they cannot flow so easily between the filaments of the solid parts, and in the case of those parts that have very irregular shapes, in the form of branches, and which consequently offer the most difficult passage of all between the filaments, they gradually become stuck there and form fat. This does not grow in the body, as flesh does, through nourishment properly speaking, but only because many of its parts join together and stick to one another, just as do the parts of dead things.

And when the humours become less abundant, they flow more easily 250 and more quickly, because the subtle matter and the spirits accompanying them have a greater force to agitate them, and this causes them little by little to pick up the parts of the fat and carry them along with them, which is how people become thin.

And as we get older, the filaments making up the solid parts tighten and stick together more closely, finally attaining such a degree of hardness that the body ceases entirely to grow and even loses its capacity for nourishment. This leads to such an imbalance between the solid and the fluid parts that age alone puts an end to life.

But in order to know more specifically in what way each part of the nutrients get to that place in the body which they are able to nourish, we must note that the blood is nothing but a mass of many small portions of food that one has ingested in order to nourish oneself, so that there can be no doubt that it is made up of parts which are significantly different from one another, as much in shape as in solidity and size. And I know of only two things that can bring it about that each of these parts proceeds to particular positions in the body rather than others.

The first is their position in relation to the route that the parts follow; the other, the size and shape of the pores where they – or rather the bodies to which they are attached – enter. For to suppose that there are in each part of the body faculties that choose and guide the particles of 251 nutrient to where they are appropriate, is to make claims to an account which is both incomprehensible and chimerical, and to attribute more intelligence to these than even our soul has: for our soul does not know in any way what they would need to know.

Now the size and shape of these pores is evidently enough to secure that the parts of the blood which have a certain bulk and shape enter some places in the body rather than others. For as we observe sieves with an array of holes, which can separate round grains from long ones, and the finest from the largest, so there is no doubt that the blood, pushed by the heart through the arteries, finds many pores in them, through which some of its constituent parts can pass, but not others.

But their position in relation to the route of the blood through the arteries is also required in order to make sure that among those of its parts that have the same shape and bulk, but not the same solidity, the more solid go to particular places, rather than to others. And it is above all on their location that the production of animal spirits depends.

For it must be noted that all the blood that comes from the heart in the aorta is pushed in a straight line towards the brain. But it cannot all go there (because the branches of the aorta which extend this far, namely those called the 'carotid', are very narrow compared to the opening of the heart from whence they come), and only those of its parts which, being solid, are also the most active, and those most agitated by the heat of the 252 heart, go there. Because of this, they have a greater force than the others to follow their course to the brain. At the entry to the brain, in the small branches of these carotids, and also particularly in the gland that physicians have supposed only serves to receive the phlegm, those parts that are small enough to pass through the pores of this gland are filtered through, and these make up the animal spirits. Those that are a little larger attach themselves to the roots of the filaments that make up the brain, but as for those that are largest of all, they pass from the arteries into the veins to which they are joined and, retaining the form of blood, return to the heart.

[Part 4. The bodily parts that are formed in the seed]

A still more perfect knowledge of how all the parts of the body are nourished is to be had when we consider how they were originally formed from the seed. Until now, I have been unwilling to put my views on this topic into writing, because I have not yet been able to make enough 253 observations to test all the thoughts I have had on the matter. Nevertheless, I cannot refrain from setting out some very general points in passing, as I hope that these are those least likely to be among those that I will have to retract later, when new observations have enlightened me further.

I specify nothing concerning the shape and the arrangement of the particles of the seed: it is enough for me to say that that of plants, being hard and solid, can have its parts arranged and situated in a particular way which cannot be altered without making them useless. But the situation in the case of seed in animals and humans is quite different, for this is quite fluid and is usually produced in the copulation between the two sexes, being, it seems, an unorganised mixture of two liquids, which act on each other like a kind of yeast, heating one another so that some of the particles acquire the same degree of agitation as fire, expanding and pressing on the others, and in this way putting them gradually into the state required for the formation of parts of the body.

And these two liquids need not be very different from one another for this purpose. For, just as we can observe how old dough can make new dough swell, and how the scum formed on beer is able to serve as yeast for making more beer, so we can easily agree that the seeds of the two sexes, when mixed together, serve as yeast to one another.

Now I believe that the first thing that happens in this mixture of seed, 254 and which makes all the drops cease to resemble one another, is that the heat generated there – which acts in the same way as does new wine when it ferments, or as hay which is stored before it is dry – causes some of the particles to collect in a part of the space containing them, and then makes them expand, pressing against the others. This is how the heart begins to be formed.

Then, because these tiny parts, which have been thus expanded, tend to continue in their movement in a straight line, and the heart, which has now begun to form, resists them, they slowly move away and make their way to the area where the brain stem will later be formed, in the process displacing others which move around in a circle to occupy the place vacated by them in the heart. After the brief time needed for them to collect in the heart, these in turn expand and move away, following the same path as the former. This results in some of the former group which are still in the same position – together with others that have moved in from elsewhere to take the place of those that have left in the meantime – moving into the heart. And it is in this expansion, which occurs thus in a repeated way, that the beating of the heart, or the pulse, consists.

But it should be noted, in connection with this material that passes into the heart, that the violent agitation of the heat that makes it expand not only causes some of the particles to move apart and become separated, but also some others to gather, pressing and bumping against one another and 255 dividing into many extremely tiny branches which remain so close to one another that only the finest matter (which I called the 'first element' in the *Principles of Philosophy*) can occupy the spaces remaining around them. And the particles that, in leaving the heart, join together with one another in this way, never leave the circuit by which they return to it, in contrast to the many other particles that penetrate the mass of seed more easily, and from the seed new particles continue to move towards the heart, until it is all used up.

And this – as those who know my explanation of the nature of light in my *Dioptrics* and *Principles of Philosophy*, and the nature of colours in my *Meteors*, will easily understand – is why the blood of all animals is red. For I showed there that what makes us see light is simply the pressure exerted by matter of the second element, which I explained was made up of many little corpuscles all touching one another; and that we can observe two motions in these corpuscles: one, that by which they follow a straight line towards our eyes, which gives us the sensation of light; the other, that by

256 which, at the same time, they turn about their own centres. And if the speed at which they turn is much less than that of their rectilinear motion, the body from which they come appears blue, whereas if they turn very much more quickly, it looks red to us. But the only kind of body that could make them turn faster is one whose tiny parts have branches so delicate and so close to one another that the only matter turning around them is that of the first element, and I have shown blood to be like this. The little corpuscles of the second element encounter, on the surface of the blood, this first-element matter, which continually passes with a very rapid oblique motion from one of these pores to the next, thus moving in the opposite direction to the corpuscles, and they are forced by this first-element matter to turn around their centres, and even to turn more with a greater speed than could be caused in any other way, since the first element surpasses all others in speed.

It is for much the same reason that iron, when it is hot, and coals, when they are burning, appear red: for then many of their pores are filled only with the first element. But because these pores are not as small as those of blood, the shade of red is different from that of blood.

As soon as the heart begins to form in this way, the rarefied blood which leaves it takes it course in a straight line in the direction in which it is freest to move, which is the region where the brain will later be formed;

257 and the path taken by the blood begins to form the upper part of the aorta. Now because of the resistance offered by the parts of the seed that it

encounters, the blood does not travel very far in a straight line before it is pushed back towards the heart along the same path by which it came. But it cannot go back down this path because the way is blocked by the new blood that the heart is producing. This forces it to return somewhat to the side opposite to that of the new material entering the heart, and it is on this side, where the spine will later develop, that it makes its way towards the region where the parts that will serve for generation will be formed, and the path that it takes in its descent is the lower part of the aorta. But because parts of the seed also press on it from this side, they resist the movement of the blood, and because the heart continually sends new blood to the top and bottom of this artery, this blood is forced to take a circular path back towards the heart, via the side furthest from the spine, where the chest will later develop. And the path that the blood takes in returning thus to the heart is what we afterwards call the vena cava.

I would not add anything here concerning the formation of the heart, if it had only a single ventricle like that of fish; but because there are two ventricles in all animals that respire, it is necessary that I try to say how the second is formed.

I have already distinguished between two kinds of parts in the portion of the seed that expands in the heart, before it takes any nourishment from 258 outside, namely, those that move apart and are easily separated, and those that join together and attach themselves to one another.

Now although both kinds of parts are found in the blood of all animals, it should nevertheless be noted that there are many fewer of those that move apart and are easily separated in the blood of those animals that have only a single ventricle in the heart than in those animals that have two. Consequently, one can conclude that there are some of these small parts that expand easily, which I call here 'aerial' particles, which are the cause of the second ventricle of the heart, and these, after the animal has been formed, are to be found inclined towards the right side.

But at the beginning of its formation, I believe that the first ventricle, which is subsequently inclined towards the left side, is rightly located in the middle of the body, and that the blood that leaves from this left ventricle runs first towards the place where the brain is formed, then from there to the opposite spot, where the generative parts are formed, and that in descending from the brain to there, they pass principally between the heart and the place where the spinal column is formed, and after that, as much from the top as from the bottom, they return to the heart.

And I also believe that, as soon as this blood comes near to the heart, it expands partially before entering the left ventricle, and because this expansion pushes the matter surrounding it, it forms the second ventricle.

259 I say that it expands because there are several aerial particles in it which facilitate this expansion, and which were not able to break loose as quickly as the others; but I say that it only dilates partially, because the portion of the seed which is joined to it, since it left the left ventricle, does not expand so readily as those of its parts that have already been rarefied there. This is why the expansion of this portion of the seed is postponed until it has entered the left ventricle, to where a part of the blood that has already been rarefied in the right ventricle returns.

And when this blood leaves the right ventricle, those of its particles that are the most agitated and the most energetic enter the aorta; but the others, which are in part the largest and heaviest, and in part also the most aerial and the softest, begin, in separating, to make up the lung. For some of the most aerial remain there, and form tiny passages, which afterwards will be the branches of the artery whose extremity is the throat or the windpipe, through which the respiratory air enters; and the largest will return to the left ventricle of the heart. And the path by which they leave the right ventricle is what will later become the pulmonary artery; and that by which they come from there into the left ventricle is what will later become the pulmonary vein.

I would add here also a word concerning the particles that I have called 'aerial', for by that term I do not understand all those that are separated

260 from one another, but only those of this number that, without being very agitated or very hard, each have their own motion, which makes the bodies where they are remain rare and not easily condensed. And because those that make up the air are, for the most part, of such a nature, I call them 'aerial'.

But there are others, more energetic and finer, like those of brandy, and *aqua fortis*, or of smelling salts, and many other kinds of thing, which cause the blood to expand and do not prevent it from condensing promptly afterwards. Many of these are doubtless found in the blood of fish, as well as in that of land animals, and even perhaps in larger quantities: this makes it possible for the least heat to rarefy them.

And the most energetic and finest parts, that is, those which are very subtle, as well as very solid and very agitated, which I shall hereafter call 'spirits', do not come to a standstill at the beginning of the formation of the lung, as do the majority of the aerial particles; but because they have more force they go further, and pass from the right ventricle of the heart via a passage in the pulmonary artery as far as the aorta.

Moreover, since it is the aerial particles of the seed that are the cause of the formation of a second ventricle in the heart, what prevents a third being formed is that, following the second ventricle, a lung is formed in which the majority of aerial particles come to a standstill.

While the blood coming from the right ventricle is beginning to form 261 the lung, that leaving the left is also beginning to form other parts, the very first of which, after the heart, being the brain. For one must realise that, while the largest parts of the blood leaving the heart go directly in a straight line to the spot in the seed where the lower parts of the head are subsequently formed, the finer ones, which make up the spirits, proceed a little further, and get to the spot where the brain will subsequently be. Next after this, as the blood is reflected back and takes its course down through the aorta, the spirits make their way a little higher on the same side near the spot where the medulla and the spinal column will subsequently be. This occurs because the movement of the blood in the part of the aorta which descends from the heart, to which they are close, agitates the neighbouring seed, and this facilitates their path towards the former side.

Nevertheless, it does not facilitate it to such an extent that they encounter no resistance at all there, which is why they also try to move towards the other sides. And in this way, while these spirits are advancing towards the spinal column, running up and down the length of it and from there pouring into all the other spots in the seed, those of its particles that excel above the others in some quality are separated from the body and turn right and left to the base of the brain, and towards the front, where they begin to form the sense organs.

I say that they turn towards the base of the brain, because they are 262 reflected off its upper part. And I say that they turn right and left because the space in the middle is occupied by those that have meanwhile come from the heart, and from there make a path to the spinal column, which explains why all the sense organs are double.

But so that we might know the cause of their diversity, and of everything peculiar to each of them, it should be noted that the only thing that can make these particles of spirit separate and take their course left and right towards the front of the head is their extreme smallness or extreme weight, or their having shapes that retard or facilitate their movement. And I observe only one notable difference among those that are extremely small, namely that some – those that I called 'aerial' above – have very irregular and impeding shapes, whereas others have shapes that are more regular and slippery, so that they are more suited to making up fine materials such as brandy or smelling salts than air.

And in examining the properties of the aerial particles, it is easily established that it is these that must follow the lowest path of all, that closest to the front of the head, where they begin to form the organs of smell; just as it is those having the most regular and slippery figures that, flowing
263 above the aerial ones, proceed by turning towards the front of the head, where they begin to form the eyes.

I also observe only one notable difference between the larger particles of spirits, which is that some have shapes which are not really as obstructive as those of the aerial ones (for, because of their size, they will have mixed very little with spirits), but nevertheless irregular and unequal, which brings it about that they cannot move one after the other but, being surrounded by fine matter, they follow its agitation; and thus having more force than any of the others, because they are more massive, they leave the middle of the brain by a shorter route, and head towards the ears, where, taking away with them some aerial particles, they begin to form the organs of hearing. The others, on the contrary, have regular and slippery shapes, which is why they act together so easily in moving one after the other, just like the particles of water, and consequently their motion is slower than that of the rest of the spirits, which means that they descend through the base of the brain towards the tongue, the throat, and the palate, where they prepare the way for the nerves needed to make up the organs of taste.

As well as these four notable differences – which result in certain particles of spirits leaving their body and in this way beginning to form the organs of smell, of vision, of hearing, and of taste – I note that the
264 others separate gradually as they find pores in the seed through which they can pass; and it is not necessary for this that there be any differences between them, only that those that collide closest to the pores enter them, while the others run their course together along the spinal column, until they too encounter other pores through which they run into all the interior parts of the seed, and trace passages of nerves there which are used by the sense of touch.

Moreover, so that the knowledge that one has of the shape of already-formed animals does not prevent one from conceiving of that which they have at the beginning of their formation, the seed must be considered as a mass in which the first thing to be formed is the heart, and around it is, on the one side the vena cava, and on the other the aorta, which are joined at their two ends, so that the end towards which the openings of the heart are turned marks the side where the head will be, and the other marks that of the internal parts. After this, the spirits have moved a little higher than the blood towards the head where, being collected in some quantity, they have taken their course gradually along the artery, and as close to the surface of the seed as their force is able to carry them; and while they followed this course, their small parts have been able to pass through all the other paths that are easier for them than those where they are. But they have not found any such paths above the spinal column, because the whole body of spirits withdraws towards there, to the extent that its force 265 allows it; nor have they found one directly below, because the aorta is there; so they have only flowed to the right and the left, towards the internal parts of the seed.

Except only that at the exit from the head, they have been able to withdraw a little inside and outside, because the marrow of the spinal column is less bulky than the head, and they are able to find some space in the former. And this is the reason why the nerves that leave the two first junctures of the spinal column have a different origin from the others.

Now I say that the spirits, which prepare the way for the nerves in the seed, have taken their course there towards the internal parts alone, because the external ones, being pressed against the surface of the womb, did not have any passages free to receive them, but they did find enough free ones towards the front of the head. This is why, before leaving them, some became separated from others without their being of a different nature, and traced the path of the nerves that lead to the muscles of the eyes, the temples, and other neighbouring spots, and then also the paths of the nerves that go to the gums, the stomach, the intestines, the heart, and to the membranes of other internal parts that are subsequently formed.

For all that, the spirits that flowed outside the head found pores on both sides of the length of the spinal column, and by these means they distinguished its joints, and became widely distributed all around the mass of the seed, now no longer round but oblong, because the force of 266 the blood and spirits that have passed through the heart to the head have of necessity stretched it more in that direction than in the other. And it remains here only to note that the last place in the seed at which the spirits can arrive in following their course in this way is that where the navel must be: I shall speak of this when the time comes.

But order demands that, after having described the course of the spirits, I explain also how the arteries and the veins spread out their branches in all parts of the seed.

The more blood is produced in the heart, the greater the force with which it expands, and in this way it advances further. And it can only advance thus towards the places where there are parts of the seed that are disposed to make room for them, and then towards the heart via the vein joined to the artery by which this blood arrives, because they cannot take any other route than this. Two new small branches are thus formed, one in the vein and the other in the artery whose extremities are joined, and which go together to occupy the place vacated by these small parts of the seed. This makes the branches that have already been formed stretch out as far as this, for unless this happened their extremities would separate. And this occurs above all because all the tiny parts of the seed are suited to flowing thus towards the heart – or at least, if there are some that are not suited, they are easily pushed back towards the surface, so that there are none below this surface in the area where the spirits spread out 267 that do not in their turn proceed to the heart. And this is the reason why the veins and the arteries extend their branches equally there in every direction.

And the truth of this should not be doubted, although one does not usually see as many arteries as veins in the bodies of animals. The reason that the latter appear so much more numerous than the former is that the blood usually comes to rest in the small veins as well as in the large ones even after the animal is dead, because the whole membrane around them contracts almost uniformly. The blood in the arteries, on the other hand, never comes to rest in their small branches, for being pushed by the diastole, it moves quickly in the veins, otherwise it falls back into the largest arteries at the moment of systole, because their tubes remain open; and thus their smallest branches cannot be seen, any more than can the white veins, called 'lacteous veins', which Aselli discovered a short time ago in the mesentery, where one would only observe them if one opened up a living animal some time after it had eaten.

268 We can yet consider here more particularly the distribution of the

principal veins and arteries, because it relies on what has already been said concerning the movement of the blood and the spirits. Thus the first agitation of the heart, which had still only begun to form, caused the tiny parts of the seed closest to it to flow to the openings in its ventricles. And by these means were formed what are called the 'coronary' arteries and veins, because they completely surround it like a garland [Lat: *corona*]. And one should be surprised to find that it has not often been noticed that there is only one coronary vein, even though there are two arteries: for this single vein can have enough branches for it to be joined to all the ends of the branches of the two arteries. And it is not surprising that the tiny parts of the seed which come from all around the heart have taken their course towards a single spot in order to enter its right ventricle, at the same time as the blood leaving the left ventricle has taken its course through two different places in order to occupy the spot vacated by them.

When the expanded blood in the heart has left it, and has taken its course in a straight line, it first pushes a large enough portion of the seed a little further than it was, towards the top of the womb, and by these means the other parts of the seed below this portion have been forced to descend towards the sides, which has brought it about that those towards the sides flowed from there towards the heart. And thus these large veins and arteries, which nourish the arms of humans, or the front legs of brute 269 animals, or finally the wings of birds, have begun to form.

What is more, the portion of the seed from which the head will be formed, pushed thus by the blood that comes from the heart, is made a little more solid at its surface than inside it, because it has been squeezed on the one side by the blood which pushes it, and on all the others by the rest of the seed which it pushes: which is why this blood cannot at first penetrate as far as the centre; and the spirits alone enter there, where they form the space in the head in the way already explained.

On this matter, it must be noted that these spirits having taken their course from the middle of the head towards three different sides, namely towards the back where they trace the spinal column, and also via the shoulder towards the left and right front sides, the matter whose place they took has been able to be drawn towards the top of the skull, in the three spaces that the three sides mark out; and from there taking its course through the two sides of the spinal column towards the heart, it makes room for the three principal branches of the great 'triangular vessel' that is between the folds of membrane that envelop the brain, and

which has the characteristic that it brings together the functions of the artery and the vein. For the matter that was in that place, being pushed by the spirits, leaves there so easily and quickly that the branches of the arteries joined to the branches of the veins through which it flows towards the heart, are merged with these in forming this vessel, which afterwards
270 extends its tiny channels on all sides inside the skull, so that it alone provides almost all the nourishment to the brain.

Nevertheless, the blood in the principal tube of the aorta, which comes in a straight line from the heart, cannot penetrate the base of the head at first, because the tiny parts of the seed are too closely packed there and exactly below a spot where afterwards a gland will be formed, which physicians have supposed serves only to receive the pituita from the brain. It exerts itself everywhere against the small parts of the seed, which resist it, and gradually drives some out, which flow from the side towards the veins sufficiently distant from there. By these means are formed those tiny branches of the arteries called the *Rets admirabilus*, which are more easily observed in animals than in humans, and which seem not to be joined to the veins.

Next, it was also raised higher towards the top of the head, in the neighbourhood of the spot through which the spirits enter the head, around which it has made innumerable tiny channels, which are so many tiny arteries that have begun to form the small membrane called the *infundibulum*, and then that which covers the duct of the ventricle that is behind the brain, and also the small tissues called the 'choroid' tissues, which are in the two cavities in front; and after that, being collected around the spot where the small gland called the 'pineal gland' will be formed, they entered all together the middle of the triangular vessel which nourishes the brain.

271 I do not need to explain in detail anything more about the formation of the other veins and arteries, because I see nothing in particular of note, and they are all produced by this general cause, namely, that when some small part of the seed goes towards the heart, the tiny channel that it makes in going there is a vein, and that made by the blood, coming from the heart in order to take its place, is an artery; so that, when these tiny channels are slightly separated from one another, the vein and the artery seem separated, because the ends of the arteries are not seen.

And in this initial stage, several different causes can make these tiny channels turn, or make one divide into two, or two collect into one, which results in the difference that one sees between the distribution of veins and that of arteries. But this does not prevent them always retaining the same connection between the ends of their branches, because the blood which passes continually through these branches maintains it.

Moreover, the branches through which this connection is made are found in all places in the body and not only in their extremities, for even if one cuts one's foot or one's hand, one does not thereby impede the blood in the leg, or in the arm.

I will add here just three examples of the division, the growth, and the joining of these tiny channels. There was no doubt at the beginning only a single tube, which carried the spirits in a straight line from the heart to the brain, but the tracheal artery, through which the respiratory air passes, is formed later (so I shall say more in its proper place), and the air 272 that it contains having more force to rise following this straight line than does the blood that comes from the heart, this tube came to be divided into two branches, namely, what are called the 'carotid' arteries.

The two veins called the 'spermatic' veins were embedded in the vena cava, each as low as the other, at the time of their first formation, but the agitation of the aorta, when the liver and the vena cava are turned to the right side, is the reason why the spot where the left spermatic vein was embedded is raised gradually as far as the emulgent vessel while that on the right remains unchanged; just as, on the other hand, as a result of the same cause, the vein called the 'adipose', of the left kidney, is raised from the emulgent vessel, where it was first, to the trunk of the vena cava, while the expansion of the liver causes the right one to be lowered. I mean what I say when I tell you that this is something I have long sought, and indeed something in which I had the least hope of success, although it has not stopped others.

The arteries and the veins that descend in mammals have a very different origin from those that are called 'epigastric', which come from the bottom up towards the abdomen. Nevertheless, several of their branches are joined vein to vein, and artery to artery, towards the navel. This happens because the former spot is the last from which the parts of the seed run towards the heart, because they have a longer route to traverse to arrive there; and because having done exactly this, the blood – 273 as much in rising through the veins in mammals as in descending through the epigastrics – which comes from one part or another through the arteries which accompany them, drives out the parts of the seed which are between the two, until it has gradually pushed them all through the very tiny passages in the veins, and in this way the principal branches of the arteries find themselves joined to the opposite arteries, and those of the veins to veins.

[Part 5. On the formation of the solid parts]

These veins and arteries in mammals, together with the epigastrics, seem to be the last of the internal parts of the seed to be formed before the external parts, and because of this the blood from the womb comes through the navel to the heart. For the agitation of the spirits causes the parts of the seed which are at the places where they pass, rather than the others, to go to the heart. And because they pass from the brain through the spinal column to several sides at the same time, they finally come together again in the same place, which is that where the navel will be formed. But before I pause to describe this, I shall explain how the heart, the brain, the flesh of the muscles, and the majority of skins or membranes come to be formed, because this does not depend on the nourishment that the animal being formed receives from the womb.

274   When the arteries and the veins begin to be formed, they still have no membrane covering them, and are just tiny channels of blood spreading this way and that in the seed. But in order to understand how their skins are formed, and subsequently the other solid parts, it should be noted that I have already distinguished above between the particles of blood that rarefaction in the heart separates from one another, and those that the same action joins together, squeezing and crumpling them in such a way that several small branches are formed around them which easily attach one to the other.

Now the first are so fluid that they do not seem to be able to enter into the composition of those parts of the body which harden; but except for the spirits that go to the brain, and which are formed and made up from the finest, all the others should just be considered as vapours or serosities of the blood, from which they are continually issued, via all the pores they find along the arteries and the veins through which they pass. Thus there only remain the other particles of blood (those that make it appear red) which properly serve to make up and nourish the solid parts; nonetheless, they do not serve this role while they are severally joined together, but only when they have come apart from one another, for in going backwards and forwards several times through the heart, their branches gradually break, and finally are separated by the same action that had joined them.

Then, because they are less readily moved than the other particles of blood, and because some branches usually remain, they come to a halt 275 against the surface of the passages through which they pass, and thus they begin to form their skins.

Then, those that come after these membranes have begun to form are joined to the first, not indiscriminately in every direction, but only from the side where, without preventing the flow of serosities, there can be vapours, and also other finer matter, namely the first two elements that I described in my *Principles of Philosophy*, which run incessantly through the pores of these membranes; and gradually joining themselves to each other, they form the tiny filaments of which I said above all solid parts were made.

And it is notable that all the filaments have their roots along the arteries, and not along the veins. Because of this, I even doubt whether the membranes of the veins form immediately from the blood that they contain, or whether they are formed rather from the tiny filaments that come from neighbouring arteries; for what contributes most to the formation of these tiny filaments is, first, the action by which the blood comes from the heart towards the arteries, which inflates their membranes, and dilates or contracts their pores at intervals, which does not happen in the veins. Second, there is the flow of liquid matter which leaves the arteries through the pores in their skins, in order to enter all the other places in the body, where it causes these tiny filaments gradually to advance; and flowing from all sides around them, it also causes their tiny parts to adjust to one another, join together, and refine themselves. But although some fluid parts can leave the veins in the same way, I believe 276 nonetheless that often it enters the other places in the body from those fluid parts which, leaving the arteries, do not take their course towards the surface of the body, but towards the veins, where they mix a second time with the blood.

And the only thing that leads me to believe that the blood of the veins contributes anything to the production of their covering membranes is that these skins are browner, or less white, than those of the arteries. For what causes the whiteness of the latter is the force with which the fluid matter flows around their small filaments, which breaks all the small branches of the particles of which they are composed, which I said above was the reason why the blood appeared red. And because this force is not so great in the veins, where the blood does not come in such an impulsive way, so that it does not make them inflate in jolts, as it does in the arteries, the tiny parts of this blood, which attach themselves to their covering membranes, still retain some of the tiny branches that make them red. But they make these skins blackish, not red, because the action of the fire that has agitated them has ceased: just as one observes that soot is always black, and that coals, which are red when they are alight, become black when they have been extinguished.

Now since the solid parts of the tiny filaments are composed, turned, folded, and intertwined in various ways, following the various routes of fluid and fine matters which surround them, and following the shapes of 277 the places where they encounter one another, if one had a good knowledge of all the parts of the seed of some species of a particular animal, man for example, one could deduce from this alone, by entirely certain and mathematical arguments, every shape and structure of each of its bodily parts. And reciprocally, if one knows all the peculiarities of this structure, one can deduce what it is the seed of. But because I am considering only the production of the animal in general here, and to the extent that there is a need to explain how all its parts are formed, grow, and are nourished, I shall continue just to explain the formation of the principal bodily parts.

I said above that the heart began to form from some of the tiny parts of the seed being squeezed by others that had expanded due to heat. But to know how it enlarges and becomes perfected, it must be borne in mind that the blood that produces this first dilation returns a second time to expand in the same place and contains in it some particles composed of several of those of the seed that are joined together and, consequently, are much larger, but it also has several that are finer, just as I said, and some of these finest ones penetrate the pores of the compressed seed that has begun to form the heart, and some of the larger ones are brought to a halt against it, and gradually drive it out of there, beginning to form 278 tiny filaments there, similar to those that I have said form along all the arteries, except only that they are harder and stronger there than elsewhere because the force of the expansion of the blood in the heart is very much stronger. Nevertheless, it is not noticeably stronger than in the first branches of the artery called the 'coronary' branches, because they surround the whole heart. This is why the small filaments that form along these coronaries blend easily with those that have their roots in the ventricles of the heart. And as these latter make up the internal parts, those that carry the nourishment from the coronaries make up the external ones, while the branches of the veins that accompany them carry back to the heart the particles of blood which are not suitable for nourishing them.

There are still many different things to consider here, the first being the way in which certain very bulky fibres are formed, taking the form of cords, which are of the same substance as the rest of its flesh. To achieve this result we must consider its ventricles to have had, from the beginning, very irregular shapes, because, the parts of the blood that they contain being unequal, they took different paths in expanding; as a result of this, they made a number of holes in the parts of the seed that they compressed, and all of these holes gradually widened, finally making a single ventricle; and the parts of the seed that separated them, having been gradually driven out from their places by the tiny filaments that make up the flesh of the heart, also made up these fibres in the form of columns.

The same thing is responsible for the production of the *valvules*, the 279 little flaps of skin that close the entrances to the vena cava and the pulmonary vein. For since the blood descends into the heart through these two entrances, stretching them as it returns and causing them to expand, the other blood that follows it through these same entrances prevents it returning via these; this is why its parts spread all around the seed that makes up the heart and makes a number of small holes there. Then the tiny filaments of the flesh of the heart drive out the parts of seed that are all around these holes and put themselves there instead, arranging themselves in such a way as to compose the valvules and the fibres attached to them. For in considering the action of the blood that descends into the heart by means of these entrances, and that which tends to leave them via neighbouring ones, one sees that, following the rules of mechanics, the fibres of the heart, which are found between these two actions, must have spread out in the form of flaps of skin and thus taken the shape that these valvules have.

But those at the entrances of the pulmonary artery and the aorta are not produced in the same way, for they are outside the heart, and only make up the skin of the arteries, which has been folded and moved along from the inside, on the one hand by the action of the blood leaving the heart, and on the other by the resistance of the blood that is already

contained in these arteries and which withdraws towards their circumference, finally making a passage through it.

And this holds generally for the production of the valvules in the
280 rest of the body. Because of this, passages must be formed everywhere,
through which flows matter which encounters other matter that resists it
in some places, but which cannot for all that interrupt its flow; for this
resistance makes the skin of the passage fold in, by these means forming
a valvule. This can be observed in the intestines, at the spot where the
excrement already collected is in the habit of resisting the flow of that
coming down; it can also be seen in the passages of the gall, and still more
evidently in the veins, at the spots where the weight of the blood that
carries it to the extremities of the legs, arms, and other parts, often resists
its ordinary course, which carries it from these extremities to the heart.
Consequently, one cannot find it strange hereafter if I say that the spirits
also form valvules in the nerves, and in the entries and exits of these
muscles, even though their small size prevents them from being observed
by our senses.

Another thing which it seems to me must be considered here is what
the heat of the heart consists in, and how its movement is produced, for
particularly since it does not cease to beat throughout its life, it seems
that all its fibres must be made so flexible by this movement that this
flexibility could easily be returned to them by an external force when it is
dead and cooled. Yet on the contrary we observe that it remains rigid in
this state, in the shape that it had previously in systole – that is, between
two of its beats – without it being easy to give it back the shape it had in
diastole – that is, the time when it beats against the chest. The reason for
this is that this movement of the diastole has from the beginning been
281 caused by heat, or the action of fire, which, following what I explained in
my *Principles of Philosophy*, could not have consisted in anything else but
the matter of the first element driving out that of the second element from
around some parts of the seed, having communicated its agitation to
them. And by these means, these parts of the seed, in expanding,
squeezed the others that have begun to form in the heart. And at the same
time some also entered forcefully into the pores between the others that
were forming the heart, by means of which they changed their position
slightly and began the motion of the diastole and after that the systole,
when this position was resumed, and these parts of seed which had been
agitated by fire, went out again from the pores in the flesh of the heart and

returned to its ventricles. Encountering other particles of seed, and on account of the blood descending there, they were mixed in with this blood, and drove out the second element from around many of these particles, and by these means passed their agitation on to them, all this blood expanding, and in expanding it sent once more some of its particles, surrounded exclusively with first-element matter, into the pores of the flesh of the heart, that is to say, between its fibres, which causes for a second time the motion of the diastole. And I do not know of any other fire or any other heat in the heart other than the agitation of the particles of blood, nor of any other cause which can serve to maintain this fire except only that, when most of the blood leaves the heart at the time 282 of diastole, those of its particles which remain there enter into the flesh, where they find pores arranged in such a way, and fibres agitated in such a way, that there is only matter of the first element surrounding them; and at systole these pores change shape because the heart lengthens, which makes the particles of blood, which remained there as if they were to serve as yeast, leave there with a great speed, and in this way entering easily ɪ... the new blood coming into the heart, they make its particles separate from one another, and in separating thus they acquire the form of fire.

Now while the fibres of the heart are agitated by the heat of the fire, they are arranged so as to open and close their pores alternately, so as to produce the movements of diastole and systole. For even after the heart has been taken out of the body of the animal and cut into pieces, provided it is still warm, it requires only very few vapours from the blood, taking the opportunity to enter its pores, to compel the movement of diastole; but when it is already cold, the shape of its pores, which depends on the agitation of the first element, has changed, so that the vapours of the blood no longer enter them, and because its fibres are rigid and hard, they are no longer so easy to bend.

We may still consider here the causes of the shape of the heart, for they are easy to deduce from the way in which it is formed. And the first peculiarity that I note consists in the difference that exists between the 283 two ventricles, which clearly shows that they have been formed at different times, and this is the reason why the left ventricle is much longer and more pointed than the right. The second is that the flesh covering this left ventricle is very much thicker at the sides of the heart than at its point, the reason for this being that the action of the blood which expands in this ventricle spreads out all around, and strikes the sides with more force than at the point, because they are closer to its centre and are opposite one another. The point on the other hand is only opposite the opening of the aorta which, receiving the blood easily, prevents it offering too much hindrance to this point, and for the same reason the heart becomes shorter and rounder in its diastole than in its systole.

I see nothing else notable here, except the skin called the 'pericardium' which surrounds the heart. But because the cause that produces the pericardium is no different from that which forms all the other skins or membranes, and generally all the surfaces that mark out the different parts of animals, it will be easier to cover all these at the same time.

There are surfaces that form first with the bodies whose boundaries they mark out, and others that form afterwards, because the body is separated from another of which it was previously a part. Of the first kind is the external surface of the skin called the 'after-birth', which envelops 284 the child before it is born; likewise, there are the surfaces of the lung, of the liver, of the spleen, the kidneys, and of all the glands. But those of the heart, the pericardium, of all the muscles, and even of all the skin of our body, are of the second kind.

What makes the first kind of skin form is this: when a body that is not liquid is produced from the joining together of the small parts of some fluid, like those I have mentioned, some of its parts will necessarily be external to others, and these exteriors must be arranged in a different way from the interiors because they touch a body of a different nature (that is to say, one whose small parts have another shape, or are arranged or move in a different way) than those of which they are composed, for if this were not the case, they would combine with one another, and there would be no surface distinguishing the two bodies.

Thus as soon as the seed begins to gather and acquire a structure, those of its parts that touch the womb, as well as some others that are very close, are forced by this contact to turn, to arrange themselves, and to join with one another in a different way from the manner in which those further away turn, arrange themselves, and join together. In this way these parts of the seed that are closer to the womb begin to form a skin which has to enclose the whole offspring; but it only achieves this some time after-wards when all the internal parts of the seed have already been driven towards the heart through the arteries and through the veins which take 285 their place, and finally these arteries and these veins also go towards the exteriors, which flow through the veins towards the heart, to the extent that the arteries advance and produce many small filaments from whose tissue this skin is composed.

As for those surfaces which are formed from a body dividing into two, they cannot have any cause other than this division. And in general all divisions are caused by this alone: one part of the body that is dividing is carried by its movement towards one side while the other part that is joined to it remains where it is, or is carried by its movement towards another side; for this is the only way in which they can be separated.

Thus the parts of the seed making up the heart at the beginning are joined to those that make up the pericardium and the sides, so that, all together, they form only one body. But the expansion of the blood in the cavities of the heart moved the matter that surrounded this cavity, in a way other than that which elongated it slightly, and at the same time the animal spirits descending from the brain through the spinal column towards the sides of the body also moved the matter near the sides in a different way. In this way, the matter between the two, not being able to follow the two different motions together, began gradually to come apart from the sides and from the heart and then began to form the pericardium. Next, to the extent to which the parts of the seed which made it up flowed towards the heart, the arteries of the different places through which they passed sent tiny filaments in their place, which, joining one to the other, formed the skin of which it is made. And what then made this skin hard 286 enough is that, on the one hand, many parts of the blood which dilated in the heart entered into the whole of its flesh, and collected there between it and the pericardium, no longer being able to proceed any further, because, on the other hand, many vapours from the blood contained in the lungs also left, to the extent to which they began to grow, and these were collected between the same pericardium and the sides. And so these vapours, compressing it from one side and another, made the fibres very hard and caused there to be always some space between it and the heart, which is filled only with these vapours, one part of which is condensed in the form of water, the other remaining in the form of air.

Index

acids, 11, 100, 104, 120, 139, 190
'action', viii, 4, 6, 20, 29, 30, 32, 35, 51, 53, 58, 62–7, 69–70, 75
'aerial' particles, 190–2
ageing, 170, 184–5
air, xv, xvi, 12, 16–21, 27–8, 139, 205
alchemy, 142
anatomy, ix, xii, xxviii, 99, 101, 140, 170–8
anger, 163
animals, viii, ix, xii, 177, 181, 190–1, 195, 200
animal spirits, xxiv, 105, 108–69 *passim*, 183, 190–2, 193, 202
aorta, 102, 173–9, 186, 190, 191, 193, 195, 197, 201
appetites, xxiv, 169
Aristotle, xviii, xxii, 6 n. 9, 8 n. 14, 9 n. 19, 19 n. 37, 26, 137 n. 42, 182
arteries, 99, 100–6, 139, 142, 152, 166, 169, 171–205 *passim*
Aselli, Gasparo, 194
astronomers, 41, 46
atoms, xvii, 12, 13 n. 29, 13 n. 30, 156 n. 51
automata, 3 n. 1, 169

Beeckman, Isaac, viii, 5 n. 4
Bitbol-Hespériès, Annie, xxiii n. 15
blood, 100–2, 120, 121, 140, 141, 167, 172–205 *passim*; movement/circulation of, xxxiv, 101, 172–82; colour of, 101, 188, 198
bones, 99, 103, 121, 171
bowels, 100, 104
brain, 99, 103, 104–6, 108–12, 117, 119, 140, 141, 142–69, 189, 191, 195–6, 205

Cano, Melchior, xxvii
causation, xviii–xix, 23, 31–2
centrifugal force, viii, xx–xxi, 35, 54

cerebral ventricles, xxv
chemistry, xii, xxiv
Clerselier, Claude 3 n. 2
Coimbra commentators, xxiii, 5 n. 4, 5 n. 5, 28 n. 55
collision, xiv, xvii–xx, 19–20, 25 n. 49, 28, 31
colour, vii, x, 85–96, 99, 101, 124, 126, 127–8, 131, 139, 149, 161, 188
comets, xi, xx, xxi, 20, 40–1, 47, 69–75
common sense, faculty of, 146, 149, 150, 169, 172
confidence, 141
Copernicanism, *see* heliocentrism
cosmology, xiii, xvi, xx, xxviii, xxix
courage, 141
crystalline humour, 125–9, 132, 138, 158, 187

Descartes, works (excluding the three works translated in full): *Compendium Musicae*, 123 n. 30; *Dioptrics* vii, xiv, xxviii, 7, 66 n. 91, 68, 76–84, 159, 188; *Discourse on Method*, vii, viii, xi; *Geometry*, vii; *Meditations*, xxviii, xxix; *Meteors*, vii, xxviii, 85–96; *Passions of the Soul*, xxviii, xxix; *Principles of Philosophy*, xxviii, 31 n. 60, 184, 188, 202; *Rules for the Direction of the Native Intelligence*, ix, xiii; 'Treatise on Metaphysics', x, xxix
desire, 141
'determination' *see* motion, direction of
digestion, xii, xxiii, 104, 141, 169, 171
dualism, xxix

ear, 122–3
earth (element), xvi, 15–21, 48, 99
Earth (planet), xx, xxi, 20, 22, 41–53, 67
elements, theory of, 11 n. 23, 16–21, 56–62, 188

206

Cambridge texts in the history of philosophy

Titles published in the series thus far